THE DRIFTERMEN
by David Butcher

Life in the tough
days of Britain's vanished
herring fleets, recalled
by the men who manned them

*Line illustrations by
Syd Brown*

Tops'l Books
9 Queen Victoria Street, Reading, England

Published 1979 by

Tops'l Books,
9, Queen Victoria Street,
Reading RG1 1SY
Berkshire, England

Printed in Great Britain by
The Creative Press (Reading) Ltd.

©**1979   David Butcher**

All rights reserved. No part of this publication may be reproduced by printing, photography, photocopying or any other means, nor stored in a retrieval system, without the express permission of the publisher.

British Library CIP Data

Butcher, David Robert
The driftermen.
1. Herring-fisheries — England — East Anglia — History — 20th century
I. Title
338.3'72'755   SH351.H5

ISBN   0 906 397   02 2 (hardback)
ISBN   0 906 397   03 0 (paperback)

Also from Tops'l Books
Sailing Fishermen In Old Photographs
Steam Fishermen In Old Photographs
Provident and the story of the Brixham smacks

Also by David Butcher
Waveney Valley
(East Anglian Magazine Ltd)

For the East Coast driftermen, wherever they sleep.

The text of this book contains many specialised local, technical and nautical words which may not be familiar to the general reader. To avoid peppering the pages with footnotes and explanations a full glossary is included at the back.

SMITH'S KNOLL LIGHTVESSEL

Contents

Chapter		Page
—	Preface	7
—	The Folk Who Told The Tale	8
1.	From Domesday To Doomsday	11
2.	The Drift Net Method	19
3.	Sail and Steam	43
4.	The Home Voyage	63
5.	Landing The Catch	79
6.	The Share System	89
7.	Life on Board	99
8.	Custom and Belief	113
9.	The Drifterman's Year	121
10.	Cousin Mackerel	133
—	The End of the Voyage	144
—	Glossary	146
—	Bibliography	152

Illustrations

All the line drawings in this book are by Syd Brown.
The photographs on pages 18, 62 (bottom) and
132 (top) are by courtesy of Jack Rose
and the one on page 112 is from the author's collection.

The portraits on pages 8 and 9 are by Rick Turrell,
with the exception of that of Jumbo Fiske which
is the copyright of Ernest Graystone.

The photographs on pages 10, 33, 41, 42, 62 (top),
80, 90, 99, 120, 132 (bottom) and 143 and on the
front cover are the copyright of Ford Jenkins.

The picture of the beatsters at the top of page 33 is by courtesy of Stanley Meades and Peter Ayres.

Front cover photograph
George Spashett, LT 184 battles her way seawards for another voyage.
Built in the 1950's for the Small Group as a drifter-trawler, she was one of the last generation of East Coast herring catchers.

Preface

This is not a book researched just from manuscripts and old records. It is living history composed of the personal testimony of the men and women who helped to make it. It has been compiled from many hours of tape recorded interviews in which East Anglian driftermen and people in auxiliary occupations called on their memories of the vanished heyday of the herring industry once so important to the British economy.

Apart from filling in background information where it was needed I have made no attempt to embroider the story. The people themselves come over wonderfully well and it has been a great privilege to set down their experiences. In doing so I have been mindful of what the 16th century French writer Montaigne once said of his work: "I have merely made up a bunch of other men's flowers and provided nothing of my own but the string to bind them."

The people who have contributed to this book represent a whole generation, for their story could be re-echoed by many others in the same age group, not only in East Anglia, but in other coastal communities too, especially the Scottish east coast and the West Country. I am aware of a bias in location because all my recording has been done in the Lowestoft area where I live, but the herring did not stick to local boundaries and neither did the men who caught them, so the story is very much a common one whether the boats were registered in Lowestoft, Newlyn or Banff.

In particular I hope no-one will take my choice of area as being a slight on our neighbouring port of Yarmouth, which was of course in its day the biggest herring port in the country. I would hate to arouse too many old rivalries! Admittedly there were distinctions between the towns where fishing was concerned, but they also had much in common and the story of one place is very much that of the other.

Before we begin I must express my sincere thanks to the folk whom I recorded and whose experiences constitute the bulk of this book. Their interest and hospitality, and that of their families also, has been a source of great pleasure to me. I hope that the time they so generously gave has resulted in an interesting book for all with a feeling for our maritime heritage. I am certain it will prove to be a valuable record for future generations, for much of what these people have to say about the social, economic and technical aspects of the fishing is not set down elsewhere. In transcribing their memories into print I have not attempted to do so in full dialect as the result would make difficult reading for 'foreigners', but I hope I have indicated enough for the reader to enjoy the richness of their speech.

There are many other people to whom I am indebted for help with the project. In particular I owe a special debt to my good friend Jack Rose whose knowledge of our local history is unrivalled. My special thanks are also due to my colleague Syd Brown for the splendid drawings which accompany the text and which could only have been done by an artist with an expert personal knowledge of fishing.

Others who deserve my grateful thanks are George Ewart Evans for his general guidance, Bob Jellicoe and Harry Eastick for information about Southwold and Yarmouth respectively. David King of the Port of Lowestoft Research Society for help in identifying some old vessels, and the regulars of the 'Royal Oak', Lowestoft who have done much to enrich my knowledge of fishing. Finally, I must thank my father for help with proof reading, Christopher Stratton of Jarrold's and Colin Elliott of Tops'l Books for their encouragement, and many other people too numerous to mention individually who have variously assisted in the setting down of this important piece of 20th century social and industrial history.

The Folk Who Told The Tale

ERNIE ARMES — born 1902 into a fishing family. Worked for 50 years on the Lowestoft market and also travelled the coasts extensively. An avid reader, especially books of local interest.

HERBERT DOY — born 1900. A fisherman through and through. Went herring catching in both world wars and also worked on Ministry research vessels in the 1950s.

ARTHUR EVANS — born 1901, the son of a leading fish merchant. Still takes an active interest in his business and also in the Lowestoft Rotary Club.

JIMMY FISHER — born 1912 in Pakefield. Came off the boats in 1935 and went longshoring. Very active since retirement in working for the Naval Patrol Service Association.

FRANK FISK — born 1899 at Wenhaston. Did all kinds of fishing during his 40 years at sea and worked for a spell at the Richards shipyard, in Lowestoft.

FLIP GARNHAM — born 1909. Began on the net store when he was about 20 and still works there mornings. A successful competitor on "What's My Line?" on BBC Television.

NED MULLENDER — born 1896 into a Pakefield fishing family. First went to sea in 1910, beginning as cook and later working up to skipper. Did patrol service in both world wars.

JACK ROSE — born 1926, the son of a Lowestoft longshoreman. Went fishing as a youngster and now works for Birds Eye. Spare time taken up by recording the history of his home town.

ANNIE SHORT — born 1902. A Lowestoft beachwoman, who worked on various net stores in her own part of the town. Interested in the breeding and showing of pedigree Burmese cats.

GEORGE STOCK — born 1903. Primarily a trawlerman, and for many years a successful skipper out of Fleetwood. Returned home after retirement and now lives in Pakefield with his wife.

JACK STURMAN — born 1891. Began work as a gardener's boy before going to sea. Became a chief engineer in 1911, after service as cook, cast-off and stoker. Remained down below thereafter.

BILLY THORPE — born 1908. Worked for several years on various kinds of fishing boat, before becoming a ship's husband. Now involved with the Patrol Service Association and the Retired Fishermen's Club.

HORACE THROWER — born 1904. Went fishing at the age of 13. Spent all his working life at sea, most of it down in the engine room. North Sea patrol service in World War Two.

JUMBO FISKE — born 1905 at Kirby Cane. Left his Norfolk farm at the age of 17 and became the best known of all driftermen. Awarded the M.B.E. for his services to fishing.

Above, *Homeland, LT125,* on her trials in 1908. Jack Sturman was on a Shetlands voyage in her when the First World War broke out and she was ordered home for minesweeping duties. Below, the *Golden Gift,* built 1910, running into port.

CHAPTER ONE

From Domesday to Doomsday

*'Of all the fish that swim the sea
The herring is the king.'*
(Traditional)

The 20th century has seen the collapse of many traditional ways of life, none more spectacular than that founded upon the North Sea herring. For many of the older inhabitants of Great Yarmouth, Gorleston and Lowestoft it must have seemed at times as if part of themselves were being torn away and stamped underfoot. In seas that once teemed with the silver shoals there is now a ban on their capture, part of an almost total restriction around our coasts and one that has been brought about by ever-dwindling stocks. If anyone had suggested 50 years ago that this would happen, not many people (if any!) would have believed such conjecture. The involvement with herrings at the time was too complete, had been going on for so long and would, as far as could be estimated, continue for ever. Even the period of depression in the 1920s and 30s, during which the whole fish trade had a bad time, held no indication of what was ultimately to befall the East Anglian herring industry. Total disappearance was unthinkable, impossible even — and yet this is what happened. And not only has a way of life passed and gone, but also the type of boat that caught the fish and many of the distinctive buildings that served the industry.

Such a change must at times seem total to those people whose lives were so closely tied up with the herring trade. It brought modest fortunes to a few, but for the great majority it meant long hours of work for little reward. Any community that grabs a living from the sea knows what it is to be hard up; and its members also have more than their fair share of sorrow. In the case of the East Anglian herring fishery, such experience was not just to be had along the seabord, for the inland villages supplied many crew members to the boats and were thus an integral part of this specialised maritime society. True, their connection with the fishing was not of such long standing as that of people living in the actual ports, but it was nevertheless of considerable importance. In fact, the combination of agriculture and fishing as an economic factor in the area's history is surely recognised in the pub name, 'Plough And Sail', which still survives in Lowestoft and which is traceable along the coastal fringe if one looks in old directories.

In purely historical terms the East Anglian herring fishery can be said to date from the late Anglo-Saxon period, with the first real documentary evidence of its importance to be seen in Domesday Book (1085-6). It is here that one notices the various annual herring rents paid by local manors to their particular lord. Then, as the Middle Ages progress, we witness Great Yarmouth's rise to pre-eminence and her attempts to control the autumn herring season at the expense of other local coastal towns and villages, especially Lowestoft. The competition between these two places over a period of 350 years would make a book all on its own, and even in our own century there was still a sense of rivalry between the respective inhabitants. It exists even now, and the planner who suggested some years ago that both towns should grow towards each other along the A12, thereby forming a conurbation to be known as 'Yartoft', obviously had no idea of local feeling!

Admittedly, by modern times it was usually a case of banter being exchanged rather than blows. Yarmouth fishermen were known as rednecks or duff-chokers, the latter term being an allusion to the light-duff or dumplings that were so important an item of the rations on board ship. The Lowestoftians were called pea bellies and this again had a dietary connection, referring as it did to the sea peas which grew wild on the local denes (and still do in places) and which were a staple food throughout the 17th and 18th centuries. There were other distinctions as well, not least being the working garb of the 19th and 20th centuries. Yarmouth men always wore blue jumpers (calico smocks), their Lowestoft counterparts tan-coloured ones — though both often wore white, undyed ones for best. Then there was a difference of technique in shooting the nets from the fishing boats themselves. The Yarmouth way was to pay the nets and ropes overboard, whereas the Lowestoft method was to throw them clear. Further variations are to be seen in the names given to certain crew members aboard the steam drifters: for instance, the second engineer was known as the fireman in Yarmouth, but in Lowestoft he was called stoker. Finally, the man whose job it was to regulate the speed of the capstan when the nets were being hauled (usually a lad in his late teens) was called a younker on board the Yarmouth boats, but was known as the cast-off in Lowestoft.

Apart from the rivalry that grew up between the two premier towns, the Middle Ages are also interesting for the amount of incidental information they afford regarding the herring and its place in the local economy. It is even possible to catch a glimpse of the men involved in fishing. There are those who made their boats and their nets important bequests to their sons; those who resented paying tithe to the parish priest; and those who broke the law in pursuit of their living. That there was wealth be to had from the sea can be seen in the number of superb churches that grace the East Anglian littoral fringe, particularly along the Suffolk coast. The perpendicular splendours of Walberswick, Southwold, Blythburgh, Covehithe, Kessingland and Lowestoft are all due in some measure to the herring. As well as the buildings themselves, a close look at their fittings (fonts, screens, tombs etc.) also reveals the fact that people were making money out of fishing. Such folk were invariably the half-and-halfers, who owned both fishing boats and farmsteads.

I suppose it would be fair to say that a Roman Catholic England must have favoured fishing more than a Protestant one, and it is significant perhaps that Elizabeth I's early governments did their best to compel people to eat fish on Fridays and Saturdays. However, such artificial stimulus was not really what was needed to boost the English fishing industry, and if any lessons were to be learned then it was the Dutch who should have been regarded as teachers. From the end of the 15th century they had organised their whole enterprise on surprisingly modern lines, employing large vessels that not only caught herrings in quantity, but which also processed catches on board. Their particular method of curing had been perfected in the late 14th century and was to become known 500 years later as the Scotch cure. It entailed removing the gill and long gut and packing the fish in barrels between layers of salt, an excellent method of preservation, which ensured quality that no rival country could match. There were around 1,000 Dutch boats fishing the North Sea in 1560, 2,000 in 1620. Many of them were substantial craft, 50 feet and more in length, crewed by anything between 10 and 15 men and boys, and with a storage capacity of between 35 and 100 lasts depending on their size. The last was 12,000 fish at this time. It grew to 13,200 later on.

All the while this great expansion was taking place, and with Dutch herring fetching two and a half times as much per last as English ones, the home industry remained small in scale. Most of the boats were little more than longshore craft, carrying a crew of three or four, and their efforts were puny compared with the busses from across the North Sea. During the reign of James I two East Anglian men, Tobias Gentleman of Yarmouth and Edward Stephens of Lowestoft, wrote pamphlets advocating the expansion and more efficient organisation of the English herring industry. Their advice, however, fell on deaf ears; and even the naval wars with Holland later on in the 17th century, which actually caused the Dutch to become less adventurous, didn't result in any great expansion. Throughout the whole of the 16th century the number of boats fishing for herring out of Lowestoft totalled about 20. By 1670 it had risen to only 25. Yarmouth, on the other hand, had around 220 vessels, though most of them were small ones.

The 18th century saw an increase in the number of East Anglian boats fishing for herring — until the last two decades, that is, when war with the American Colonies and with France brought about a serious decline in the number of boats. The eventual end of the Napoleonic Wars saw an upward turn in the fishing and this time it was to be a trend that continued right through to the end of the 19th century. There are several reasons for this, not the least initially being the decline of the Dutch, who were no longer the economic force they had once been. They still fished for herring in the North Sea, but on a smaller scale, and by about 1830 they had largely ceased to be part of the annual 'invasion' of Yarmouth. Thus ended the connection between them and the country's premier herring town, a bond that had survived two centuries of political and maritime disputes between England and Holland and that had led to a resident community of Dutch people in Yarmouth, complete with its own church and pastor. Dutch cargo boats still continued to put into Lowestoft, however, to pick up consignments of salted herrings (a practice that continued right up until World War Two), and at times the old Fishermen's *Bethel* on Commercial Road had so many of the visitors in its congregation that services were conducted in their language.

Apart from the decline of the Dutch, another factor in the gradual expansion of the East Anglian herring industry in the first part of the 19th century was the development of the three-masted *lugger*. This was a bigger, more efficient boat than the craft used hitherto and led to greatly increased catches. It was Yarmouth particularly that prospered with this vessel, for she had a harbour and (until 1830) Lowestoft didn't. Direct landings on to the beach obviously necessitated either smaller fishing vessels or the use of rowing boats to ferry catches from the luggers to the shore. In 1844 Yarmouth had 120 luggers of her own (many of them two-masted by now), plus numerous smaller craft, and around 40 cobles that came down from Yorkshire to work for local owners and merchants in the autumn season. After the railway arrived (also in 1844) the number of boats grew to 200 or so a decade later, and the increased prosperity was reflected in the building of a large wharf and herring market between 1867 and 1869.

Lowestoft's mid 19th century boost not only had to do with the railway, but also with the man who brought it from Norwich to the town. Sir Samuel Morton Peto was one of the great builder-contractors of the Victorian era and his moving into the immediate vicinity was Lowestoft's great piece of good fortune. He bought out the bankrupt harbour company in 1844 and proceeded to develop and improve the whole complex, marrying this expansion with the coming of the railway in 1847. He guaranteed to local fish merchants, at a meeting in the Lowestoft town hall, that any catches landed in the morning would be delivered 'alive' in Manchester the next day. And he kept his promise. By 1872 the number of boats fishing for herring from the port had risen from 80 to 210, and the population had grown from 4,837 to 13,620 in just 30 years!

The foundations were now well and truly laid for the golden age of the herring, which lasted from the final quarter of the 19th century until the outbreak of World War One. It was a two-phase affair, with the years 1880-1900 being the heyday of the lovely ketch-rigged *dandies* (a development from the earlier luggers) and those afterwards seeing the rise to pre-eminence of the steam drifter. Not only did the arrival of steam power bring about a large increase in catches from the local boats, it also hastened and encouraged the annual autumn visit of Scottish fishermen and curers which had been getting under way in the last two decades of the 19th century. Old men who can remember those days before the First World War still talk about the sheer quantity of herring landed and the money earned. Their tales are not romance. Those were the days of the thousand drifter fleet in Yarmouth and Gorleston, when you could walk from one side of the harbour to the other across the boats, and when Lowestoft had between 700 and 750 vessels in port.

For the sake of brevity, let us consider the incredible year of 1913, the one that topped all others. The 1,006 boats fishing out of Yarmouth (264 local, 742 Scottish) landed a total of 824,213 crans of herring (a cran weighed 28 stones), while the Lowestoft fleet of around 770 drifters (350 local, 420 Scottish) brought in nearly 535,000 crans. And on that note everyone went to war. There was little herring fishing between 1914 and 1918, and not just because of the U-boat menace around our shores — most of the drifters were on charter to the Admiralty for patrolling duties or for minesweeping. After the war things began again in a mood of great optimism, so much so that in 1920 there were 1179 boats in Yarmouth for the home fishing, 973 of them Scottish. But how quickly it all turned sour. Catches came nowhere near pre-war levels and the collapse of European currencies left many fish curers and merchants ruined — or very near to it. The thirties saw the decline continue and by 1937 Yarmouth had only 87 drifters and 17 *drifter-trawlers* registered at the port, Lowestoft 135 and 84 respectively. The number of Scottish boats was also much reduced and catches were about a third of what they had been in 1913.

With the onset of World War Two the boats were either laid up or hired out once more to the Admiralty. When they began fishing again it was the old story of the downward spiral, though perhaps this wasn't at first obvious in a sea that had had a six year rest. By the middle fifties, though, the writing was on the wall and seen clearly by most people connected with the industry, especially after the disastrous season of 1955, when the lowest quantity of herring ever recorded (up till then) was landed. By the early sixties Yarmouth had all but ceased to be a herring port (except for *longshore* craft) and is now very much geared to servicing the North Sea oil and gas industries. Lowestoft had a handful of drifters fishing up until about 1967-68, but is now entirely given over to the trawlers that had always been so important a part of her economy.

And thus the curtain came down on the North Sea herring, which was not only a fish, but a way of life as well. At one time it must have seemed that the sea would yield its silver harvest without ever stopping, but that was not to be. The quantities being landed before the government introduced its ban on herring fishing were minute compared with the huge catches that once made Yarmouth and Lowestoft the world's two great herring ports. There are various theories as to what occurred to end it all, but one factor is undoubtedly overfishing. The North Sea stocks have been decimated. The purse-seine net has snatched the swimming shoals; the close-mesh trawl and suction pipe have cleaned out the nursery grounds. Sad that so much of this 'industrial' fishing has been carried out not to feed people directly, but to provide fish meal to feed cattle to feed people. What a wasteful chain of conversion! Nothing is more nutritious than the mature herring, nothing more beautiful visually when in prime condition. Worshipped is too strong a word to use, but the fish was reverenced by many of the men who went hunting for it. As one of them once said to me: 'Our lives depended on it. Who would ever have thought that we'd live to see it disappear?'

That remark sums up admirably the attitude of the people whose experiences are recorded in this book. Their generation was born at the height of the North Sea herring industry; and they have lived long enough to see it all vanish. Many years ago someone worked out that every steam drifter afloat gave employment to a total of 100 people. Consider then the overall effect of 600 locally registered boats in Yarmouth and Lowestoft before the First World War, to say nothing of the arrival every year of around 1150 Scottish vessels. It meant that from October to December everything in the two towns was geared to herring, and even when the boats departed for other waters many people were still required at home to service the industry. In Lowestoft particularly fishing was everyone's life-blood, and for many youngsters leaving school the choice of job was quite simple: the boys went on to the boats or down on the fish market, the girls into the net factories or beating chambers.

Apart from the employment angle, the herring fishing also made its mark on Yarmouth and Lowestoft in another way, one that is still visible although the industry itself has gone. This is the pattern of urban development that took place in the late Victorian and Edwardian eras. Broadly speaking, it falls into two categories: streets of smallish two-up, two-down terraced houses, built speculatively for rent and inhabited by all kinds of working folk; streets of larger, terraced villas, with bays at the front and three or four bedrooms within. Many of these latter were bought by the successful drifter skippers, the lesser boat owners and the small fish merchants. In Lowestoft there was such a congregation of fishing boat masters in Worthing Road and Sussex Road, at the northern end of the town, that both these streets became known as Skipper Rows. It was the fashion at one time for an aspidistra in a brass pot to stand on a round table in the downstairs bay window. The room had also to contain a piano, even though it was likely that no one in the house could play it. Some of the houses even sported two instruments! Oh, the excesses that even a modest affluence can lead to!

Of course, the spread of domestic buildings wasn't solely due to the fishing, but it was the underlying factor. You soon get an idea of its former importance if you walk these streets and look at the names not only on individual houses, but also on groups of cottages. Many of them still hark back to the fishing, by boat names, by geographical place names and by general maritime references. Without having to think too hard, I can summon up Mizpah, Belvedere, Newlyn, Rathmullen, Oban, Lerwick, Compass, Mariners and Anchor. Finding these old name tablets on the fronts of houses is very rewarding for it says so much about a bygone era. Their treatment too in terms of decoration is also interesting, because some remain as they were when the houses went up, others have been painted over and a few obliterated completely.

There is another detectable legacy in the form of bricks and mortar — the buildings that once housed industries allied to the fishing. Those that survive are but a small remainder of what once stood above ground and they have invariably been converted to other uses. In Yarmouth much of the quay area has been taken over by the gas and oil companies, and also by freight services across to Holland. Offshore supply craft and container vessels now lie where the herring drifters once unloaded their catches. Various light industries have moved into many of the old net stores, and there is only one firm remaining that cures (smokes) fish on a large scale.

The story is much the same in Lowestoft, though the recent North Sea industries have made less impact there. Most of the net stores (a few remain to serve the trawling industry) are now builder's yards, paint shops, car repair centres, fibre-glass moulding works etc. One has even been changed into an indoor bowls rink, another houses an office equipment showroom and yet another manufactures carry-cots. The smokehouses too have declined, until there are only two or three left working at the present time. Most of the large, commercial ones have either gone or have been converted to other uses, while the small backyard ones often serve now as garages or garden sheds. However, their tall, narrow outline and pantiled roofs are still very much a feature of the streets in the Roman Hill area of the town.

The chief legacy of the herring fishing era, the living one, is to be found in the memories of people directly involved with the industry before it collapsed. They were so much a part of it, and it of them, that many of them still find it hard to believe that everything has gone. There is no sense of bewilderment on their part; it is more a case of acknowledging that what they had always regarded as immutable could change and did change. Those people born between about 1890 and 1910 were the members of the last generation for whom the herring was not only a livelihood, but a way of life as well. They were at the end of a tradition stretching back centuries and they lived long enough to see the end of it. More than this, they were part of the process which saw modern technological advances applied to an old, established industry and they were glad enough, at the time, to help put into operation the ideas and methods which had been pioneered by their fathers.

As it turned out, the steam boats (and the diesel craft that succeeded them after World War Two) had a relatively short stay in the total span of the herring fishing over the centuries. Once they had gone, the face of both Yarmouth and Lowestoft was drastically changed because their passing was also the passing of a staple industry. I suppose that for many of the people brought up in both towns during the herring era to see it all disappear must have produced feelings not unlike those which might be experienced by a native of Barnsley if all the coal mines were to close down. The analogy is probably imperfect, but it does serve to make some sort of comparison. No herring scales, no coal dust; no muck, no industry.

It is perhaps the boats themselves that constituted the biggest loss, for they combined utility with great visual appeal. One is reminded of this in the sheer number of postcards and photographs of steam drifters that were produced in the heyday of the fishing, and it's hard to believe at times that so many vessels disappeared so quickly. But disappear they did, the majority of them getting scrapped, while others were converted almost beyond recognition and sold off for various uses. This was what made August 9th, 1978, such a sad occasion, because it was on this day that the *Lydia Eva YH 89*, last of the East Anglian steam herring drifters, left Yarmouth for good. She had been in her home port for five years as a floating museum, but now her owners, The Maritime Trust, were calling her to London for permanent exhibition in St. Katharine's Dock. As she came down the river for the last time, leading out the Tall Ships at the start of their race to Oslo, there must have been many people present that morning who saw not only the departure of an old steamer, but also the passing of the age she represented.

A ransacker in his working gear of long smock, leather buskins and lace-up boots, posed before a typical East Coast brick and flint bond wall.

CHAPTER TWO

The Drift Net Method

'Herrings Galore,
Pray Master?
Gay Master,
Luff the little herring boat ashore.
Pray God send you 8 or 9 last —
Fair gains all.
Good weather,
Good weather,
All herrings — no dogs.
Chorus: Fair gains all!'
(Traditional — the Gorleston boys salute
to boats leaving harbour)

Herring were caught by the method of fishing known as drift netting. It was an efficient way of catching the shoals of fish that once swam so plentifully in the North Sea, but it was selective in its operation because the immature specimens passed through the meshes, leaving only the larger fish caught. Herring tend to lie on the bottom during the day, but they rise towards the surface at dusk to feed upon the plankton that constitute their diet. It was when this swimming up took place that the drift nets claimed their silver harvest of the sea. Basically speaking, the technique was very simple: all you did was to suspend a line of nets vertically in the water, with your boat drifting along behind on the tide, and let the rising fish swim into them. There was of course more to it than that. The nets themselves were a rather complicated piece of gear needing great skill to make up and handle.

It is hard to say with any degree of certainty how and when the drift net way of fishing first arose. Certainly the Dutch had developed it to a fine art by the end of the 15th century, but it must have been practised in one way or another for hundreds of years before that. The manfares and flews that are mentioned in mediaeval wills are obviously drift nets of one kind or another, and their depth is always reckoned in scores of meshes just as net depths always were right up until the end of the fishing. The material used to make the actual net itself was either hemp or flax, the latter fibre having its name perpetuated in the word lint — the term always used for the mass of meshes in each individual net no matter what it was made of. To complete the net and make it ready for use, this lint had to be fixed in a frame of hemp cord, which would enable it to hang square in the water.

Once the individual nets had been rigged, they were laced to each other down the sides to form a fleet. This fleet was attached at its top end to a long, thick cable known as a warrope, and this in turn was suspended from wooden casks or bowls which kept the whole chain of nets afloat. A line of corks along the top of the nets themselves gave added buoyancy, and the depth at which the fleet hung in the sea could be varied either by altering the length of the seizings (ropes which fixed the nets to the warrope) or by adjusting the strops (ropes which secured the warrope to the bowls). The actual nets themselves consisted of sections about 6 feet in width, and this enabled them to be made up to a particular size by varying the number of ranns, as they were called. If one takes the

FIG 1

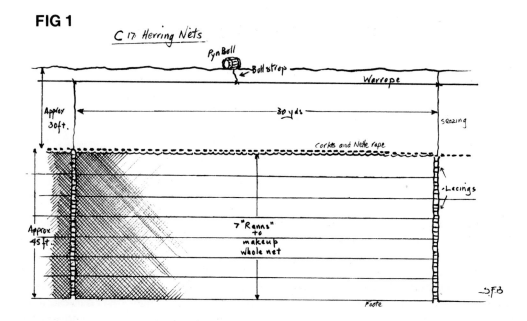

FIG 2

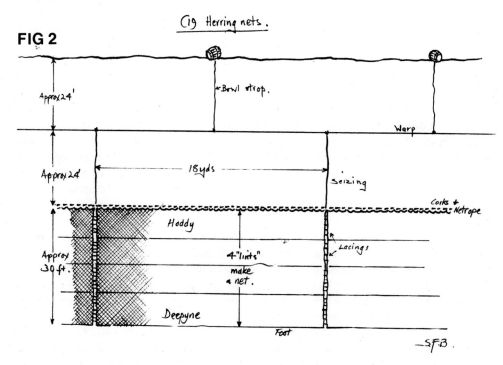

average mesh size as being something around one inch, then the rann itself was 6 score meshes deep, and this identifies those 12 score nets one reads about as 2 ranns joined together. Unfortunately nothing much is known about length, but somewhere between 15 and 30 yards (depending on the size of the boat) was probably the order of the day. The size of the particular fishing boat would also determine the number of ranns in each net, as well as the number of nets carried, of course.

Figure 1 shows a drift net of the early 17th century, as described by Edward Stephens in his pamphlet, 'Britaine's Busse'. All the requisite dimensions can be worked out from the text and one is entitled to assume, I think, that this particular kind of net was typical of those used on the larger herring boats. The principles of rigging and working don't differ a great deal in essence from those described by J.G. Nall of Yarmouth in the middle of the 19th century, except that herring nets had by this time become a good deal smaller and were suspended lower down in the water. The net(s) that Nall talks of can be seen in figure 2, and the similarities and differences between both types can easily be worked out. One notices that the nets were still being made in sections in the mid 19th century, with the topmost one known as the hoddy and the bottom one called the deepyne (deep un?)

By the time J.G. Nall wrote his excellent treatise on the East Anglian herring fishery (1866), the hand-made hempen nets he describes were already on the way out. Early in the 19th century a Scotsman, James Paterson, had invented a machine loom (there had been earlier models before his) to weave nets from cotton, and this may have been the most crucial development in drift net fishing throughout its long history. He established a factory at Musselburgh in 1820 and from then on the cotton Scotch nets gradually took over from the older variety, until by the end of the century their use was universal. The one thing that Paterson could not achieve was getting a knot on the meshes that would not slip, but once Walter Ritchie of Leith had invented a mechanical way of doing this the cotton net was supreme. It had two great advantages over the traditional twine variety: it was more efficient at catching herring (greater killing power, as many fishermen put it) and, being of a lighter and more manageable material, the boats could carry larger fleets of nets.

Once the net-making machines had been perfected, the lint (as it was still called) came off the loom in the correct depth needed to make up the individual nets. Broadly speaking, for there were exceptions to the rule, the Scotch net of the steam drifting era was 35 yards long by about 12 yards deep (18 score meshes). The lengths actually came off the loom at 55 yards, but this was reduced once the net was set within its rope framework in order to get the meshes to hang correctly in the water. Nothing was cut away or removed, but the lint was no longer stretched out flat; it had been set in by a third so as to be slack and flexible. Figures 3 and 4 show the classic Scotch net and the way it was rigged.

The meshes were boxed within a frame of cords — headings down the sides and sidecords along the top and bottom (this may seem rather confusing, but as the net came off the loom what became the top and bottom in the water constituted the sides). Once this had been done, the net rope and the back rope were secured to the net by twine norsels (or ossels); these were about 6 inches in length untied and finished at around 4 inches when fastened. There were round corks (about $2\frac{1}{2}$-$2\frac{3}{4}$ inches in diameter and about $1\frac{3}{4}$ inches deep) on the net rope, with 3 norsels in between each cork. Both the net rope and the back rope were double lengths, one with a right hand turn and the other with a left, and this was done

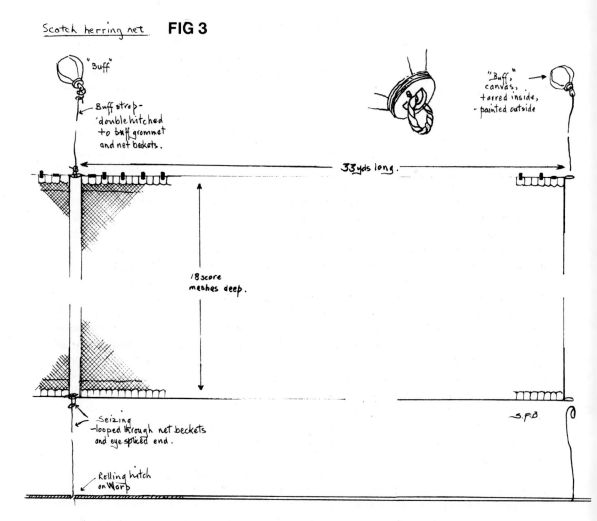

to prevent the net from rolling up on itself — chittling as it was known in the trade. The other thing they had in common was that they both ended in loops, called becketts, through which the strops and the seizings were tied. At the top and bottom of the net was an area of thicker cotton mesh, about 6-9 meshes deep (it varied), known as the oddie. This acted as a reinforcement between the sidecord and the lint proper and was probably a corruption of the word hoddy, which had been the term used for the top section of the 19th century twine net.

When it came to actually working the nets, there was one great difference between the cotton ones and their twine predecessors — the position of the main rope or warp (derived from warrope). It had always been above the nets in the days of twine because there was sufficient weight in the hemp to get the nets to hang down in the water. Cotton, however, was a much lighter fibre altogether so the warp (a manilla rope of some 3 to 4 inches in diameter) was transferred to the bottom of the line of nets, where its weight ensured that the nets themselves hung correctly. The floats used to hold up the whole system were no longer the old wooden bowls, but large, pear-shaped, canvas buffs, the size of 3 to 4 footballs, which were inflated either by a pump or by human lung-power. Figure 5 shows a steam drifter working her nets. The short length of rope called the tissot was tied on to the warp and then secured to the stem of the boat so as to take the strain of the whole fleet of nets while fishing was in progress.

There were some 600,000 meshes in each individual Scotch net and the mesh size varied according to what grounds were being fished, because the size of the herring differed. For instance, the North Shields variety was generally smaller than that caught in the autumn home season. Every full mesh that held a fish made the adjacent 8 meshes incapable of catching any and this distortion was an important factor in the fishing capability of the nets. Even with a haul of one cran per net (800-1,000 fish, depending on size), generally considered to be quite satisfactory, only about 0.14% of the meshes had been effective. This puts the claim of 'a fish in every mesh', commonly used to describe a good haul of 2-3 crans per net, in its true perspective.

Any fishing gear requires a great deal of care and maintenance, and the cotton drift nets were no exception. In order to preserve them from the rotting effect of salt water, they were first of all dressed with creosote and then immersed regularly in a solution of boiling water and cutch. This substance was the resin of the Acacia Catechu (hence cutch), an East Indian tree, and its use gave the nets a greatly extended working life. The process of preservation was known as tanning or barking, a reference to the days of the old hemp nets when an extract from oak or ash bark had been used to do the same job. It was just one of a number of maintenance tasks carried out both on board ship and, more especially, at the many net stores where the riggers and ransackers and beatsters all played their part in keeping the fishing boats at sea.

William Garnham (born 1909), known to everyone as 'Flip', spent the greater part of his working life on a net store both as rigger and ransacker. The latter job entailed checking the nets after the women beatsters had made them up from new lints or had repaired the ones already in use. It obviously referred back to the days when herring nets consisted of a number of ranns laced together and any that had been repaired were scrutinised carefully (literally, to search or seek a rann). The word beatster is also very old. It means someone

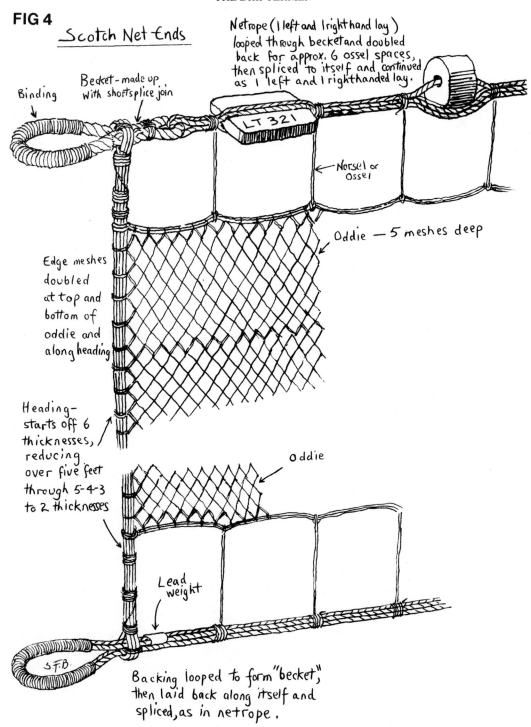

THE DRIFT NET METHOD

who repairs, makes good, and Chaucer uses the verb in 'The Reeve's Tale' when Oswald is describing the rascally miller of Trumpington: 'Pipen he koude and fisshe, and nettes bete'. Flip Garnham talks about the ransacker's job:

'To start with, you git the spoilts from the market. You know what a spoilt is — a torn net orf the drifters. They're brought up by horse and cart on to the denes. Two men then trick 'em out. That means they half the net. One take the back and the cork-line, the other one the lint. And they spread it out, and they look at these nets slowly through their hands and see what work is in the net. If thass a good net and not much work, we put a single knot in the tie. You go to the next net. If thass a little worse than what the single knot is, then we put a figure 8 in. Then we might git one with a double figure 8 — well, that mean to say when thass dry that goes to the top loft to be done by the beatsters at their leisure. But the single knots, they go on the store and are done straight away. I mean, they can git them out quick, specially when the boats are waitin' for nets. They can git them out quick. A figure 8, well they're done after the single knots are done, because they take longer. If you git too many single knots, then they go round the town to the outside beatsters, by horse and cart, or lorry, as the case might be. We had over 100 outside beatsters at one time o' day; used to take about 250 nets out. They used to git about 10 bob or 15 bob a net, and when we'd give 'em a week to do 'em we'd go and collect 'em back and then they're ransacked. The girls beat the nets and then they're passed to the ransacker. The word ransack means to look through, which we did. We undid the net after the girl had done it up and we pulled it down the store. There might be a net rope broken or a back broken — well, that all had to be repaired. Or there might be a cork out or a forelock down or the becketts want servin'. All that, thass the work of the ransacker. Well, after he's completed the net, he folds it up and puts it in the bay.

'Sometimes the nets would be right wet when you picked them up from the market. They were a hundredweight apiece, yuh know, a wet net. When you've gotta pick that up orf the market and put that on your back, the knack of it is your knee. We used to spread them out Scotch fashion on the denes there, full length. There used to be five people sometimes on the cork line and two on the back ends, and we used to pull them out flat — what we called Scotch fashion. That means the heavy net was right out tight. And when they were dirty you used to lay one on top of the other, givin' a little space each time. And if they'd had a lot o' herrin' in, they were full up wi' muck and scales and old maise (roe). We used to leave that on there and let it wash out in the rain.

'We used to rig our own nets. The nets are really 55 yards long, but when you rig a net up you take a poke. I mean, you put three norsels in your sidecord and therefore that's drawin' it up all the time. You're drawin' that 55 yards inta 35. Now the back rope was 35 yards exactly, but the cork line was 2 foot less, the simple reason bein' that the cork line is a 9 thread rope, where the back is 12 thread. Your cork line is thinner and therefore it stretch more. You see, you git a two foot stretch and that correspond with the back so everything's square. Otherwise, if one is longer than the other, you don't git a diamond shape in your mesh. And the net must hang diamond shape so the herrin's head go in the mesh.

'There was 91 osséls on the back rope and about 187 on the net rope. There was about 84 corks and they'd be done in thirds and doubles. A third mean you count 3 ossels, or norsels, and put a cork in: 3 norsels, cork; 3 norsels, cork. Then when you git towards the middle o'

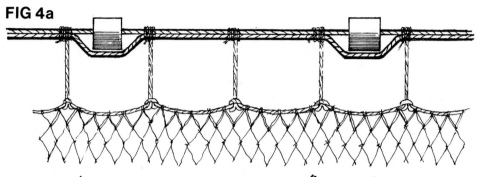

Cork & Norsel spacing – "Scotch" net

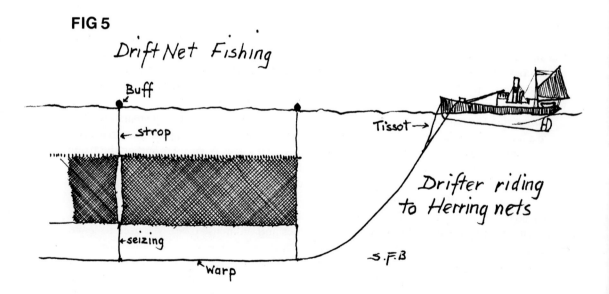

THE DRIFT NET METHOD

the net, where more weight is, you put 1, 2 norsels, cork, and right in the very middle you put 2 doubles in, double corks to take the weight. Now say that net gits t' be a year old, then we put an extra cork in the middle. You increase your corks on account o' the weight o' the net gittin' older. As a net git older, it gits heavier, therefore we hafta put in extra corks to keep it buoyant.

'Now a strop, a buff strop, was attached to the net as you know, and so was a seizing. Well, a strop was 3 fathom long and a seizing was 4 fathom. When the ships went to sea, they took 2 lots o' strops, a 3 fathom strop and a 2½ fathom strop. That mean to say they shot the 3 fathom strop, and if the herrin' were all over the net the skipper was satisfied. But if the skipper hauled and found all his herrin' in the top o' the net, then he took that 3 fathom strop orf and put the 2½ fathom strop on. That meant that'd bring the net up so they got a good spread over the whole net. A seizing was 4 stranded and a strop was 3 stranded. I suppose a strop would be about 1½ to 2 inches round and the other one about 2½ because a seizing was stronger and went on the warp. The back rope and net rope were 12 thread and 9 thread, like I said before, and the measurements were about an inch round for one and ¾ for the other.

'I reckon young Johnny Freestone and me set a record for riggin' up a net. We rigged a net up from scratch in 5 and 20 minutes. And thass goin', that is! Thass thread all your corks, serve your 2 eyes and set up all your ossels. When you rig up a net you splice 2 eyes in your net rope and thass what the beckett is fixed on to. The net rope was 2 ropes actually, a right hand turn and a left hand. When all that rope was new, we hatta tow that across the denes, take the turn out. You used to make fast to a lamp post outside ol' Arthur Gibbs's shop and tow right to the old swimming pool, one on each shoulder! And then you used to come back and coil them inta baskets — right hand and left hand. With a left hand rope and a right hand rope, you've got to coil them that way. If you try to coil a left hand rope right-handed, you'd soon bugger it up.

'I worked for J.B. Ltd. That was Jack Breach. But he was on his own then — afore he went in with Hobson's. And thass where I've been ever since. Yes, I started there in 1928; my money was 28/5d. I got married in 1935 and they give me a rise of 5 shillins. I worked at Hastings House. We had three stores: Hastings House, The Shoals and The Kittiwake Yard. Hastings House is now Seagull Engineering. Philip Oldman was our carter; I used to draw the hoss when he wun't there. Our nets were made at Beeton's, Sunrise Networks, East Street, in Low'stoft. We used to git our new nets and they were white. They were just the lint itself and we had the girls there to put the sidecords on and the headings on. They done that, and after they were done we had to tan them, three consecutive tannins. They had to be done well; they had to boil for two days easy. And then they were took on the denes and dried. Then the tan copper was got riddy again — into the copper again, boiled, out again; in again, out again. And then after that they got a thorough dry. Then they went inta the creosote pit. Inta the creosote pit they went and then out onto the denes to hang for about 10 days till they got a thorough dry. They got so dry that you could see all the crystals o' the creosote. I'm talkin' about the ol' pre-war creosote, which was the best stuff, not the stuff you git now. When that dried there used to be crystals, and weren't that terrible stuff to work with! We used to have the ol' skippers come up and give us a hand, and they used to skin. Their arms used to skin, their faces. Oh, terrible stuff, that was!

'Then we had the other job of tarrin' warps. That was a terrible job, that was! We had an old iron copper and we used to have all the warps coiled up and down. They all hatta be spliced up first naturally enough; put your long splices in, your eyes, and hell know what. And then you hatta put tar in this thing. You'd put a shute up and roll a barrel o' tar up and pour it inta this copper. Then you'd light your fire, but my God you hatta be careful! We boiled it once and that all went over. There was a rare ol' mess! There was a terrific fire out there! Once you'd got it goin' you'd pass all them warps through by hand, through the hot tar (that wuz hot enough so that'd hurt you). I used to be the nipper, what they called the nipper. That mean to say that the warps were fed into that copper and I'd pull them through. I had a bit o' sackin' wrapped round my hand so I could nip the rope. I hatta nip the tar out of it, see, so there weren't too much surplus. Then there wuz a chap stood next to me who pulled it through a trow (trough) and then last of all you had the coiler. He went up and down the denes layin' the rope in flakes (serpentine fashion) — that was what they called the ropewalk.

'The warps were 120 fathom long. When a ship had shot at sea they reckoned there was about a mile and a half, two mile o' rope out. You see, the ship used to vary in the length o' fleets they used to have. Some had 91 nets, some had 97. That was always an odd number, never an even one. The warps were joined by bend lashin'. You had a long eye about 15 foot long, see, and you put one end o' the next rope to be joined through one end of that eye. Then you kept passin' it through and round and round till you'd almost used your length. Then you back-hitched 'em to finish with. You used to tuck the last bit in with a marlin spike. If they parted, the warps, then you had to splice the two ends together. Our warps were from British Ropes and they were a 5 strand rope, about 3½ inches in diameter. They used to tow new warps out behind the boats to take the turn out. If you took it out the wrong way, you know, you ruined it, ruined the rope. Yes, if you didn't know what you were doin' with a coiled warp, you could mess it up. One skipper messed one up completely. You couldn't do nothin' with it. That went all long-jawed. That mean to say the rope is gone long in the strands so you can't do nothin' with it. You couldn't splice it, only chop it up and put it on the tan copper fire.

'In 1928 you could git a net for about £3-15s. A box o' cutch would be about 30 bob; £1-50 now, but thirty bob then. And I know in 1968 a net was £19 and a box o' cutch went up to £13. Warps were about 80 quid apiece the latter part o' the time, but early on I spose that you could git a warp for about 20 quid easy. They were terrible things to splice when they were new, I'll tell you — when they were hard. Cor blimey, we've had some times with them. Then, you see, that bloke invented that special marlin spike. If I'd have invented that, I'd have bin a millionaire. What a simple idea! Oh dear, or lor. I mean, years ago when you got on the denes and you started splicin' a warp (I mean an old warp full o' tar), you had to force your marlin spike in. I've done so much forcin' that I've had a hole in me stomach when I finished. You see, you forced your marlin spike in between the strands an' got them opened out all ready to splice. Then when you drew the spike out that all closed up again. But there was this bloke who invented this special thing. You put the spike in and that was hollow, so you could then pass the strand through the hollow spike. Lovely job! I passed up somethin' there, didn't I?

'The net stores in our firm were divided up into different rooms. There was 12 girls

THE DRIFT NET METHOD

workin' on Pevensey Castle; there was eight on Explorator; there was four on Boy Alan; there was four on Dan Frary's room; and there was three on Ben and Lucy. They were some o' the rooms we had. You see, where I worked, although that was Jack Breach Ltd., there were a lot o' small companies involved, little boat owners. Like Dan Frary, he had the *Ben and Lucy LT 714*. Lugs Seago, he had the *Boy Alan LT 331*. And who else was there? Oh yes, the *Gervais Rentoul LT 740*. You see, they all had their individual rooms and their own nets. They were the little boat owners. They had all their own nets, own buffs, own buff strops, own seizins, the lot. Even when we tanned 'em we had to make sure we kept their heaps o' nets away from the others. Say the *Victor and Mary LT 1190* come in and you sent a couple o' nets down from the *Ben and Lucy*, you'd never hear the last of it! See, you mustn't git the wrong nets for the wrong boat. If you did, and they found out, well —! Now the *Boy Alan*, he had special buffs an' all. He had the ol' canvas buffs with a bladder in. And he was the only one. Well, I was paintin' these blinkin' buffs one day. I had them hung up 'cause you had to paint the buffs, you know. And I was paintin' one when he walked in. Cor, didn't he lay inta me! "You know diff'rent to that! Paintin' that with the bladder in!" he said. "That paint'll go through on to the bladder and that'll be destroyed," he say. "That will perish. Take all the bladders out". Well, you know what happen when you take a bladder out of a case. I mean, how can you paint it? But I hatta do it.

'They were white, the buffs, but you also had your spell buffs. You had a quarter blue, which meant you painted one panel blue. Then you went to your half blue, which was two panels painted blue, and then to your three-quarter. The pole-end buff was all blue, but some of them did actually have an old Dutchman's bowl (cask) with a pole on it. They marked the quarter stages of the fleet o' nets, them coloured buffs. Then you had your monkey, the chequer bowl, which was red, white, red, white all the way round. That was five nets from the end of the fleet, so therefore when you were haulin' and saw your chequer bowl you'd say, "Thank the Lord, there's only five more nets to haul". But if you saw a blue buff in with your chequer, well you'd know that you had a foul. Yes, you had a foul there somewhere. I painted a pole-end buff green one day and I sent that down to the *Norfolk Yeoman LT 137* — poor ol' Doff Muttitt. He put a marlin spike in it! He dint like green. He wouldn't have it. There was wooden heads in the buffs one end; they acted as a kind of plug. You used to put Stockholm tar round them so that when the buffs turned upside down in the water all that tar used to go round the head and seal it. Cor, that was funny stuff! The ol' boys used to use it for piles. Thass the finest thing out for piles, Stockholm tar. Yes, you stick that up your behind an' you don't worry about piles no more!

'When I first started on the store we used to start at 6 o'clock in the mornin' an' finish at 6 o'clock in the evenin'. And durin' the Depression time we hatta work an extra hour for no extra money! Yes. There were no jobs, you see; you had no choice. There was nowhere you could go. I mean ol' Jack Breach used to swear at me an' call me all names under the sun, but I couldn't retaliate because there was nothin' else for me to do. My first job, you know, when I arrived at the store in the mornin', was to empty his muck-bin. Yes, that was my first job. I used to empty that every mornin' inta the tan copper pit and burn it. And one mornin' I forgot. Now that was his first job, when he come through his gate, to lift that lid up. And I forgot it! "Flip!" he yelled out, and o' course I come runnin'.

"Yes, guvnor?" I say.

"Have you emptied the muck-bin this mornin'?" he say.

"I forgot", I say.

"You'll forgit your bloody hid one o' these days, won't yuh?" he say. "Git the bugger emptied now!"

'There weren't a man more watched by the Gestapo, I reckon, than ol' Jack Breach was by us. You see, directly the ol' man went out of the yard that was my job to see which way he went. If he went to the market, all well and good. Then the boys used to say when I come

back, "Where's he gone?" and I'd say, "Down to the market". Well, that was what they were waitin' for. "Right, we'll git the fags out and have a smoke". You see, you weren't allowed to smoke on the store. But if he went the other way goin' to The Shoals, that was a dead loss. My job was to see which way he went, so I used to git a barrow load of cinders out of the stoke-hole and wheel it out, just to see where he'd gone. Well, one partic'lar mornin' I didn't have no cinders, so I put an empty barrel on and went chargin' out o' the gate — and there he stood! Well, o' course, I pulled up right sharp opposite and he come and looked inta this barrel. "Where the hell are you goin' with that?" he say. I hatta think right smartish, so I say, "I've spilt some tar on the floor in there, guvnor, and I'm now goin' across the sea wall after some sand to clean it up".

'Durin' the home fishin' we used to work the clock round, specially if there were any spoilts. Sometimes you'd git a net that might have a little slit in about a yard and a half long, or about a foot and a half. Well, that was a single knot straight away. A girl could work that out in about an hour, easy. But if you got one that'd been dog-eaten, bitten by dogfish, well that'd never be done. That'd take 3 or 4 days to do, at least. Well, that was a figure 8 straight away. A double figure 8 was even worse; they'd be put up the top loft. Then, when things were slack, like durin' the summer when all the boats were away at Shetland, we'd bring 'em down an' the girls could work at their leisure on them. An overhand knot was a condemned net. You put your overhand knot in the tie and you hung it up. When it was dry, you folded it up an' put it in the bay an' it went for garden-lint. They all went for garden lint, the overhand knots, sold to the net merchants in Low'stoft. A spronk was just one strand of a mesh split. A crow's foot was two; you know, like a claw. If you got a lot o' them, it was a bad net because they take longer than a split. There's been the time when they were so hard up for nets that they actually left spronks and crow's feet in the nets.

'The beatsters at home got so much a net. Very little an' all! After the Second War that was a pound a net, thass what it was. The latter part o' the time, I'm talkin' about. After the war they'd all do it in their sheds. I mean, they had decent sheds made. I'm talkin' about the beatsters what got married, you know, and then had a child or two an' couldn't git to work. Before the war they got about five bob a net. We used to take out 250 at a time. Normally you'd send all your best nets round the town. I mean, that was no good sendin' a dog-eaten net down, otherwise you'd never have got it finished. Every girl, when a net'd been finished, hatta chalk her initials on the corks. Well, I mean, if you pulled that out along the store when you were ransackin' it an' you see a couple o' spronks or crow's feet in there, you'd normally look and say, "So and so, come an' finish your net orf." But if that was someone you liked, you'd ignore it, pass it by.

'There was 31 meshes to the yard on a net. That was the average size. In fact, that was all

THE DRIFT NET METHOD

we used after the war, 31 to the yard. Before the war we had various sizes — 31's, 31½'s, 32's, 33's — but then they done away with them because a lot o' the fishin' grounds closed after the war, like Ayr and Dunmore. In 1937 I went away to Lerwick to look after some of our firm's drifters, you know, and I had a couple o' beatsters on the store there an' they used to sit down on a chair an' cut out with a pair o' scissors. That seemed so comical to me, them sittin' there cuttin' out. O' course, you've gotta trim your mesh first afore you mend it. I mean, you've got a lot o' loose ends hangin' about. Well, a girl on the store in Low'stoft would just git a knife and trim it orf. But in Scotland, at Lerwick, they'd sit there cuttin' out with a pair o' scissors, which seemed very comical to me.

'At the top and bottom of the net there was the oddie. That was thick mesh about six meshes deep. You see, that was the lint between your net and your sidecord. Then on your sidecord you'd have your ossels. On the back rope there would be eight knots (on the meshes) and then an ossel; on the net rope five knots and an ossel. The heading was down the side. That was four or five strands of hemp, or stuff like hemp anyway, and that was the heading. Ossels was the correct word, but most people said norsels. I often wonder how that got to be norsels. I mean, all the time I was on the store I said "norsels". But when I was on "What's My Line?" an' they said (when I'd done my mime) "What was that?" I said "Settin' up an ossel." And they told me afterwards that sounded so comical; so broad, you know, there on the television — "Ossel".

'In 1932, when things were very, very slack, three or four of us got sacked. That weren't really the sack; we got stood orf. Well, rather than stand about, I went to sea on the *Herrin' Gull LT 330*. I went three-quarter shareman; went away to Lerwick. We were away 13 weeks and our gross earnins were £878 — for 13 weeks! Our charges (expenses) were £421. We paid orf at £14 a share. I was three-quarter shareman and I took up 11/3d. Mind yuh, my mother had bin drawin' 15/- a week allotment all the time I was away. She was a-drawin' and I was a-livin' orf me stockie. When we paid orf ol' Jack Cunford, the skipper, come to me and say, "What about it, Flip? Are you goin' to sign on for the fishin'?"

I say, "Jack, I'd rather pick up blinkin' herrin' orf the market." But as it turned out ol' Jack Breach see me on the market. He see me standin' about, so he say, "You want to go back and git your job back, don't yuh?"

I say, "Thass what I'm waitin' for, guvnor."

"All right then," he say. "Go and git started agin."

They dint all git stood orf that time; there was just three or four of us. I'd never bin on the drifters before. I'd bin smackin' and I'd bin trawlin', as a boy, but I'd never bin driftin'. But, you see, beins I knew all the knowledge o' the nets, I went. Like Jack Cunford said, "Well, you've worked on a store a good many years. You know what there is to know."

'Afore the war all our nets were 18 score deep. But when we come back in 1946, started agin, they altered 'em to 17 score. And we kept 17 all the time then. There used to be a net what we called a pearl twist; that was a two-stranded cotton, but that was very, very hard. If you got a fleet o' them on a drifter they used to be piled way up. They used to catch herrin' like the devil, but the only trouble about 'em was they used to lose a lot. They were so hard that the herrin' used to drop out. They'd fish marvellous, but they used to lose a lot. So they done away with them and went in for what we called Egyptian cotton, which was a softer cotton altogether. And then we even had nylons, but they weren't a success. Yeah, the

THE DRIFTERMEN

Harold Cartwright LT 231 had a whole fleet o' nylon nets, and he went to Aberdeen with a 70 cran shot and they nearly turned 'em down on the market there. They said they were dog-bitten, but they weren't. What'd happened, you see, was that the net was so sharp that it'd cut the heads orf the herrin' as the crew were scuddin' them out. After that he only had about 20 nylons in his fleet and the rest were cotton.

'Some o' the boats, if they'd bin out in rough weather, used to lose half a fleet o' nets or even three parts. Well, when they come in, they'd all got to be replenished. You'd got to git fresh nets aboard and a new lot o' warps as well. Sometimes we've even bin down to git the herrin' out of a boat! That was when the crew'd bin haulin' a long time, like 14 hours, even 20 hours sometimes if things were bad. The blokes were tired when they got in, so they used to send up the store and tell a storeman to go down the market and help git the herrin' out. And clean the nets as well. Git the pilchards and scads out. They were terrible things they were, scads (horse mackerel). They'd soon poison your hands. I solved that problem, though, when I used to do it. I used to pull the hids; break the hids orf and let them drop. I've bin down the market scores o' times gittin' herrin' out, doin' scads and pilchards, pullin' up nets. Pilchards I dint mind, but scads were terrible. Best preserver there is for a net, you know, pilchards. The *Norfolk Yeoman* had a whole fleet o' nets once full o' pilchards and, do you know, them nets outlasted any nets we had. And we used to know. Directly you boiled 'em you smelt the old oil in them. Pilchards. They used to catch them out on The Knoll, all over.

'In 1953 our firm had a tan copper built at Richard Irvin's yard in Aberdeen. The simple reason was, you see, we had diesel ships then and therefore they either had to go alongside another steamboat to git the steam for the tannin' tank, or tan ashore. Well, our firm thought that'd be beneficial if they had a tan copper built in Aberdeen, and I got the job o' goin' away to do the tannin'. I used to go away about the end o' March and be back about the end o' August. Well, when the boats go away in March, thass what they call the first voyage. They take all the old nets on that. Well, an old net only want tannin' about once a month, so they can work them a month without worryin' about tannin'. Therefore on that voyage I was tannin' the boats once a month.

'Well, after 12 weeks them ships went back agin to Low'stoft. They'd finished that first voyage. They were home about 10 days and then the next time they come to Aberdeen they'd all got bran' new nets in. Well, that was my job then to cure them nets. I mean to say, they've never bin in the water before, so they'd got to be cured. They had to be shot for 10 shootin' nights, which mean to say they had to be in the water 10 nights and then come ashore to be tanned. And they had to be done for three minutes; boiled for three minutes; no more, no less. The object o' tannin' a new net is to boil the bacteria out o' the knots and put the tan in. If you just put a net in the tan copper in lukewarm stuff and take it out, you might as well not do it at all. You must boil; but you mustn't boil too much on a new net 'cause you boil your dressin' out — thass your creosote. You boil that out and the nets go white, and then they go rotten. Well, thass your first tannin', after 10 nights. Then the nets are put back on board and laid in the kid, all night, till the next day, when they go to sea. That give 'em a chance to dry. That was no good them goin' to sea with wet nets 'cause when they were chucked over the side all the tan'd come out. And that meant all that work was in vain.

Above, Beatsters on the North Denes at Lowestoft in 1904. Net drying frames are behind them and the lints being repaired are hooked up on iron stakes driven into the ground. The girls are not in their working clothes. They put on Sunday best when they heard the photographer was coming. Below, Scots fishergirls hold up a baulk of split fish prior to kippering in the smoke house.

THE DRIFTERMEN

'The next time the nets go 14 shootin' nights. Then they come back and I had to do them the same. Then they go 21 nights, which is three weeks, and then after that the month. Then they're cured; they're finished; they're a lovely job. They're cured and they'd go a month every time after that. But that is a bit of a problem to cure a new net if you don't know what you're doin'. If you tan 'em more'n three minutes, if you give 'em five, you can kill a net. Or if you don't do 'em long enough. I remember Joe Thompson o' the *Lord Hood LT 215* when he had some new nets. He say t' me, "Flip, are you goin' to tan my nets for me?"

I say, "Yis, Joe, I'll do 'em for yuh."

He say, "You know how long they're gotta be, don't yuh?"

I say, "Yeah, three minutes."

He say, "Make sure an' give 'em three minutes."

I say, "All right, Joe. Blast, I've done it afore, haven't I?"

Do you know, he brought an alarm clock with him! He stood it on the tan copper and made sure he got his three minutes! That was an outside tannin'. He weren't with our firm, but I used to do other boats as well.

'When you're tannin' you fill your tank right up to the top. You go right up to the top of the iron tank. Then you've three or four rows o' bricks up above that where your lids go on. Well, you put your water in up to the top of your tank and then, if you were doin' an old net, you'd put a box o' cutch in. That was a hundredweight. You'd chop that inta two bits and put it in two baskets, put them on your derrick, heave them up and lower them inta the copper. Now you'd keep them on the derrick, see, so they were suspended in the water. Well, as you boil, you see, that gradually melt and drip, drip, drip through the bottom o' the baskets and then you create your cutch proper. Well, after thass all melted away, you have to bale most o' that liquid out inta a tank — what we call the strengthenin'. So, therefore, when you do a dippin' o' nets, say 27 nets, and you take them out, you find your stuff in the copper's gone down about six or seven inches. Well, thass when you use the stuff out o' your tank. You put that back in to keep the strength up. You have to keep the strength up all the time. See, if you're tannin' a dry net, you'd be surprised how much stuff go out o' that copper. But if you're doin' a wet net, then that actually increase what's in the copper. We used to tan the nets just as they come in from the sea. If a boat wanted tannin' quick, we dint wait for 'em to dry. They used to go inta the copper wet. Course, then you had to give 'em a helluva tannin' because that used to cool down the cutch so much. You'd use a lot more cutch on a new net. When you tanned a new net you wanted 2½ boxes for one dippin'. We had a copper down The Shoals — what we used to call the big copper — and that used to hold 100 nets. That was all right puttin' them in, but that was a bugger gettin' them out! Normally a copper would hold about 24. We had a little copper at The Shoals which held 24, but the one at Hastings House held 27. The one at Aberdeen, that held 24.

'When you'd got your nets in the copper you put two gratings on the top and two heavy boards across them. On top o' them two boards you put a trunk or a wooden box, and on top o' the wooden box you put a chock. Now in the wall beside the copper there was a hole, an aperture, made inta the wall and steel-lined, where you stick a big pole in. You've got to lock that in there and have that go across the top o' your chock and your box. Then you put a rope on the end o' that pole, which go down to a block and tackle thing on the ground.

THE DRIFT NET METHOD

Then you go down and you pull down on that till the nets are right under the tannin'. They'd boil over otherwise, you see. You hafta hold them down somehow. Another thing is if you chuck cutch in loose. Cor, you'd have a mess! You'd git it all in amongst your nets, and once that dried, you see, you'd had it.

'The tan coppers were made o' steel. You'd be surprised the heat you had to establish to git the ol' copper boilin'. You know, from cold. We used to have Welsh Hard, we did. Used to go and cart our own years ago. Used to come up there by the railway sidin' and we used to go and git the ol' cart and go down and git it ourselves. We had two coppers at Hastings House — one inside the gate and one between the stores. Well, we dint use the one agin the gate unless that was absolutely necessary because when the wind was in the nor'ad that used to blow on the ol' man's house. You see, his house was in the yard there. And once we were a bit pushed, we were tannin' a lot, and our foreman say to me, "Flip, go and light that tan copper agin the gate — the little copper."

I say, "The wind is blowin' right, Victor."

He say, "Never mind about that. I want it."

So anyhow I lit it and stoked it up well. And o' course the smoke belched out, and that belched all over the house, you know. Out come the maid. Could I stop the smoke? I say, "Blast, I can't light a fire without smoke!" And o' course they told the ol' man when he come home, ol' Jack Breach. So he come tearin' out after me.

"What the hell are you a-doin' with that smoke?" he say.

I say, "What the hell do you want me to do? Whitewash the coal?"

Cor, he was a rum ol' boy, he was!

'When the boats made up at the end of a voyage, everything come up the store. Everything. And everything had to be looked at. All the jobs had got to be done for the next voyage. Everything. Buffs painted, all the riggin' re-served, and the blocks had got to have the pins knocked out and black-leaded and put back in again. Our pub used to be the 'East of England'. That was our pub, specially on a Saturday when we were waitin' for the boats to come orf the North Sea. We dint leave orf, you know; we used t' wait. Say the *Margaret Hide LT 746* was due up at half-past 12, well rather than go home and git our dinner and come back agin, we used to carry on workin'. Well, at twenty past 12 probably you'd look down the Beach and you couldn't see nothin' o' the *Margaret Hide*, or anyone else for that matter, so we used to go in the pub and have a couple o' pints. And then you'd look through the winders and see her go past!

'The mate used to give all the orders for a boat. Everything was left to the mate. Whatever was wanted. If he wanted 10 nets and three or four buffs, or a bundle o' strops and seizins, he used to tell the carter or the ship's husband. Normally the ship's husband, and he'd pass it on to the carter, and he'd tell you when he got up the store. But I had an ol' wireless set on the store; the ol' trawler band, you know, and I used to git all the information orf that after the war. Ol' Jumbo Fiske and Blackeye Soanes, they used to shout on, "Flip! Want two ton o' ice, four nets, bundle o' strops an' seizins, an' four buffs." And o' course I used t' git the ol' motor, and away I used t' go up the ice comp'ny and have all that riddy on the market when they come in. I got wrong though. I had the G.P.O. after me. I used t' have the trawler band on, you see, full bore. And o' course you know what language they used t' use! Well, an ol' girl livin' up above Hastings House, up there in the High Street, she heard it. And she told the G.P.O. and they come down and played hell up. The bloke say, "You aren't sposed to have that on." So I say, "Well, what's it on the blinkin' wireless set for if you ean't use it?" I used t' pick them boats up at Aberdeen on that ol' Vidor. In fact, that got so common, you know, that ol' Victor Bond, the foreman, used t' come across in the mornin' and say, "Have you had n' news this mornin'?" So I'd say, "Ol' Jumbo's on his way in with so many cran. He want two ton o' ice and four nets." Yis, that was how we used t' go on.'

There speaks the ransacker. Annie Short (born 1902) was a beatster and she too has very clear memories of her working life. She too worked for Jack Breach at one time. Born and raised on the old Lowestoft Beach Village, her story is that of many girls of her generation and background, and she tells it with great gusto and humour. Her father was Albert Spurgeon, the famous Lowestoft lifeboat cox'n, a renowned humorist and practical joker. His daughter obviously inherited much of his sense of fun:

'We used to start at 8 o'clock on the stores and finish at 5 o'clock with an hour for dinner. And durin' the fishin' time, the home fishin', we used to work overtime till 9 o'clock and have half an hour for tea. When you started apprentice you had to do all the odd jobs — you know, like scrubbin' the toilet out and doin' the errands and makin' the tea, and all that sort o' thing. You had a year apprentice and a year improver. You got a shillin' a week apprentice and half-a-crown improver, and at the end o' that time you were a beatster.

'I started work in 1916. I was 13 when I left school and there wun't much t' do 'cept beatin' and braidin'. And I tell you, we never got a sight o' money; we never did! Our beatin' store was right out on the denes and, oh that was a cold place! That was Charlie Day's place where I learnt my time and I went t' Breach's afterwards 'cause you got stood orf, you see, in the summer time. They didn't want you and you had to git where you could. When I first started at Charlie Day's they had a double piece o' cotton in the corners o' the nets. But they weren't no good; they used to pull the nets to pieces. And then down at the Shetlands they used to git a lot o' these dogfish, and one time I remember they brought these nets back and they were torn t' shreds. Bran' new nets! Well, we done what we could; we took two or three nets and made one up out of 'em. That was when I was on Charlie Day's store, where I learnt my time. Then I went t' Breach's, and then the latter part o' the time I worked for ol' Germany Greengrass.

'Now a Scotch net is 55 yards long and 18 score deep, new. And when you went apprentice you used to do all what they called the spronks. That was just one strand of a mesh broke. Then you went on to a crow's foot, which was two strands broke. And then they'd gradually put you on to splittin' the knots and cuttin' the holes out. You done a year apprentice and a year improver. When you were improver, you'd be put with a full beatster t' learn all the ins and outs, and then when you'd done that time you were on your own. After a time you'd be good enough t' go on to makin' up nets from new lints. You know, puttin' the lints inta the sidecords and headins. We were on piecework on that job; we could git five shillins and odd a net. We used to give the apprentices a shillin' each at the end o' the week for fillin' our needles and puttin' the norsels on. There used to be one norsel to five meshes on the cork rope and one to eight on the back rope. We had wooden needles then. The workhouse men used to make them and they used to charge us 2d for them — big wooden needles. We used t' use the small bone needles as well.

'If you got a net what was split from one end to the other, from clew t' herrin' as we used

THE DRIFT NET METHOD

to say, we used to what we call temporary join it. We used t' join up t' the headin' if that was possible and take praps two or three mashes (meshes) out either side o' the split in order t' clear the rags. If you understand what I mean. You know, like you would with a dress; take a piece out. But o' course when you joined all that up you had t' git the norsels right, do the net wun't hang prop'ly. Oh yes, you had t' git that right. There was some work hung t' that, I tell you. Well, you imagine a big sheet, a bedsheet, and thass ripped right across — where do you start?

'And then if we got a dog-eaten net, oh that used to take your time! You see, the dogfish used to bite the net. If there was a herrin' in it, they used to take a piece right out. I have bin an hour takin' up just one little piece. But we used t' nearly always have 2 of us on a net, on a spoilt. A needle o' cotton would go nowhere! If we got a bad net we worked 2 on a net t' git it done. We used t' git 15/- a week and 3d on a net. Every net we done we used t' git 3d. And more often than not, when we were on Breach's store, we used to mark the nets down and the forewoman would keep count and then we'd share out what we had earned afterwards. There was several of us worked there, and that din't matter whether you'd done four or five nets and one girl had only done one — we used t' share. Course, 3d on a net wun't much, you know, but every little helped.

'I was workin' at Breach's the latter end o' the First World War, at The Shoals. We were workin' from 8 in the mornin' till 9 at night durin' the fishin', with an hour for dinner and a half hour for tea. We used t' git Thursday nights orf and finish one o'clock Saturdays. My money used t' be 18/5d a week, with what I earned extra on the nets — and thass the truth! And as for ol' Jack Breach, he was as uncouth as he could be. Gret fat ow man. A gret fat ow lump, with a gret big pod. And I remember when he used t' come up on the store he used t' shake the floor! He was that heavy. We used t' wear these navy blue, striped, cotton pinafores. We had t' provide our own; they didn't give 'em to yuh. And one o' the girls come one mornin' without a pinafore and Jack Breach come and stood in front of her. And I shall never forgit his words! He say, "Hulloo, gal, what're yuh forgot yuh airpron?"

'When we got the nets new they used t' come in a sheet and the forewoman used t' cut them out. There'd be so many of you put the cords on, the sides and the headins. The ransackers used t' put the cork rope and the back rope on and sometimes, if the apprentices wun't busy, they used t' git so much for puttin' the norsels on. And they used t' fill our needles up an' all. We used t' do that as well, fill the needles — git all the needles we could and fill up dinner times. When you were headin' you got all the lint together in your hand in a thickness. Then you'd tie that together and put it on a hook on the wall, and then you'd put your heading across. You'd keep it straight as you were workin', o' course, and so you would on the sides. The sides were the same, but you couldn't always gather all of the side together to tie, because that was longer, so you'd praps git half of it. That had t' be straight though. If that wasn't, everything would be out o' true.

'Sometimes when the boats come in we found coppers in the cork rope what the fishermen had put in so they could reckon that was lucky. You know, pennies cut inta the corks. And then you'd git nets what were jellied; they'd had jellyfish in them. And do you know, I've sneezed with them! They'd dried out and they'd go up your nose like pepper. And I've sneezed and sneezed till my nose ha' bled. And your eyes would water. Oh, they were terrible things! When they used t' bring the nets in and we used t' go run 'em out on the denes, they used t' tan them sometimes if they thought they were full o' jellies. They would tan them t' take that out. That didn't use to do the trick always; you'd still sneeze!

'When the boats used t' come up in August and they were a bit pushed, they'd run the nets out and we'd go mend the big holes out on the denes. You'd come out o' the store, and we used t' have an iron thing about three or four foot high with a hook on the top and a spike on the bottom. And we used t' work that inta the ground t' hang the net on. And we used t' git a bottle o' drink and 2d or 3d a day extra for bein' out on the denes. That was when the boats come up in August. They used t' run up about the end o' August and they'd be at home about a week afore they went away on the North Sea. And you used t' do these nets for them t' take back with them. But didn't they use t' be bad when we really got them up to the hook!

'I remember one time when we were on the denes mendin' nets and there was a boat bin across t' Ymuiden. Course, they used t' run across there a lot. And when they'd bin across t' Ymuiden they had t' fly a flag, a red flag with a white ball in the middle. That was t' let the customs know they'd bin across the other side. And we were out one day on the denes and there was a party o' visitors round us. And one of 'em said, she say, "Whass that ship flyin' that red flag for?" And I say, "Thass when they're got red herrin' aboard." Well, laugh! Them girls, they dint know what t' do with theirselves. And the questions they used t' ask us, the visitors! They were innocent really. They used t' ask us if we worked out there all the winter; things like that, you know. I used t' hate the denes, what we called mendin' out on the denes. You didn't all go, but if you did you'd git so sunburnt. And you'd git very tired as well. They used t' run the nets one over the top o' the other and there'd be about four of us on a net. We'd run one o' the ropes, the net rope or the back, over t' the other side and then gather the nets up and the men'd fold them. Or else praps we would fold them and the men would tie them up and carry them. We were never allowed t' carry a net alone. Well, we couldn't; there'd allus have to be two of us.

'I dint mind workin' on the store. I worked at Hastings House, on what they called the Pevensey Castle. There was about a dozen of us on there. Well, o' course, we never had no wireless so we used t' have a sing-song. And Jack Breach's wife, she was one o' these jumped-up clients. You know, she thought herself above everyone. And she was only a beatster! Any road, she used t' look down on us, and she had two skivvies. Well, one day she sent one o' the skivvies up with some lard — for us t' grease our throats with when we were singin'. What about that? I don't say we were posh singers, but that used t' pass the time away. And then if one o' the girls had bin t' the pictures or had read a book, she'd tell that like a story t' the rest. Oh yes, while you worked. Course, you weren't allowed t' sit down, you know; we were standin' all the time. You couldn't sit down t' do them nets. It was impossible. The only time you got t' sit down was when you sat down t' fill your needle. And if you were too long doin' it, you were told t' git on with it. From the forewoman!

'When we worked on ol' Germany's place, ol' Germany Greengrass's — that was a little ol' place at the back o' East Street — Dan Elliott, the foreman, he done some cotton and he done it with some tar sort o' stuff. And there was only me and this other woman on there 'cause the boss's daughter dint work at night; she wun't old enough. And that was a bitter cold night that night. That was just afore Christmas. And I said t' her about makin' up the fire 'cause we'd got a little ol' stove up this corner, and that weren't a very big room. So she

THE DRIFT NET METHOD

say, "Well, we can't put no more coal on." But she got this cotton and we chucked that on t' the fire. There was a little left over, you see. And she got a poker and held it up and that set the chimney alight! And they went down the market after ol' Germany, but there wun't nothin' there when he come back. We'd got some water and put it out. But oh, that did stink. And that blazed. Went up the chimney like billy-o!

'When I was workin' there for Germany we used t' hev a cup o' drink in the mornin', a cup o' cocoa. No milk in it. We used to pay 3d a week for this cocoa and then we never had a break in the afternoon. We used t' work 8 till 1 o'clock in the mornin', but we wun't sposed t' work longer than five hours without a break, so we used t' git a break from quarter to 11 till 11 o'clock. And come tea time, there used t' be a little ol' shop across the road which sold whilks and that sort o' thing, so we'd have some o' them. And do you know, I used to enjoy the home fishing. We were busy, but I enjoyed it. You could see our winder from the road, the beatsters' winder, and the person I worked with she say t' me one day, she say, "I've got some socks t' finish for Bill for Chris'mas." So I say, "Well, I've got some knittin' I want t' do as well." So we used t' git the net on the corner o' our little finger, and we'd be a-knittin' away and o' course the net was a-jumpin' an' all. Well, anybody who was goin' along the road could see the net a-workin' through that winder. I'd praps have half an hour and then she'd have half an hour. And that was how we got our knittin' done!

'Now I'm goin' t' tell you somethin' about my husband's sister. She was a bitch! Not as regards bein' nasty, but she was a proper one for a game. She worked on a store at the end o' Carlton Road, near the 'Fighting Cocks' public house. Well, there was a little tiny winder there from what I can make out and the boss used t' look through there, keep an eye on things like. He was a decent boss, though; not like some of 'em. And they had an apprentice work there and she never wore no knickers. So my sister-in-law up-ended her one day and did the necessary. The boss went in and told his wife. He say, "Ol' Nellie, she's a-shinin' away!" She black-leaded this girl's behind, my sister-in-law! And the girl went home, never washed, went t' bed; and o' course her mother went up the store. They had t' hide my sister-in-law up. But that was the sort o' game she would play. Yeah, she black-leaded this girl's bum 'cause she never wore no knickers!

'We had a boy work with us up at Charlie Day's who was a bit short of a shillin'. Coo, we used t' have some games with him! That store used t' stand right out on the denes and we used t' git t' know the smacks comin' out of the harbour by the patches in the sails. Anyway, this boy's mother was beautiful and clean, and he used t' bring his dinner what she'd made. She allus made him meat pies with a knobble on the top. Well, we used t' pinch the knobble orf it and he used t' call me "Bloody Lizzie" for doin' that. And ol' Cutts, the foreman — he was a rum ol' boy — come inta work one day with this gold stripe. Durin' the First World War, if the men'd bin wounded, they used t' wear a gold stripe on the sleeve o' their coat. And ol' Cutts brought this inta work and we sewed it on t' the behind o' this boy's coat! And he went home with it on! Oh, we used t' have some laughs.

'At the time I told you about when I was workin' for 18 and odd a week for Jack Breach, overtime was included in that 18/5d. But after the war was over, a year or two after I think, they got us a Board o' Trade wage. And even then that was only just 21 shillins a week. But we didn't git anything on a net then, not once they'd brought that in. Mind you, Breach's did use t' give us a bonus. If we done so many nets, we used t' git a bonus, and we used t' git that every quarter or half-year. Sometimes, when we worked, we used t' have a separate net up t' the one what we'd worked on durin' the day, and we used t' git 5/6d a net on that. That was after tea. We used t' work from half-past 5 to half-past 8. That was mendin', that was, but when we worked on Germany's place we got so much an hour. I can't tell you exactly how much, except that wun't a lot. What we used t' do with our overtime money was put it in Tuttle's club t' git our clothes with. T' git a coat. Praps you'd put two shillins a week in, or half-a-crown. And you'd git enough t' buy yourself a coat 'cause you could git a decent coat in them days for thirty bob. And you'd have that once every two years; you didn't have a new coat every year. You'd have your workin' coat and then you'd have what you called your goin' out coat. One would be for weekdays and one for Sundays. You never wore the same coat Sundays as you did durin' the week.

'Beatin' and braidin' was about all there was for girls when I was young. The braiders (people who made up trawl nets) only used t' work six weeks apprentice, but they didn't git a penny durin' that time. They didn't git no money at all, but their time was only six weeks. Beatsters used t' git a shillin' a week apprentice, and then when you'd finished that first year that went up t' half-a-crown. And then that gradually went up till you got your full money. If you were any good at the job you got your full money after a bit, and that was 15/- a week afore extras. I remember when there was a depression in 1921. The boats wun't earnin' any money or anything, and they dropped the weekly wage t' 13/-. So that wasn't good money; you couldn't call it good money. In the winter time, fishin' time, after I was married, I used t' have nets at home when the kids were little t' git some extra money. I done 'em in the scullery; bin up here till 12 o'clock, one in the mornin', t' git 'em done. That was a 5/6d a net and you were lucky if you could do two in a week. And if you had what they called a dog-eaten net you got 10 bob. And that'd take you a fortnight. At one time I did garden lint and o' course thass only a lot of old crap. That was when I was at home and I got 1/3d for doin' a piece o' net 100 yards long and a yard wide! You had t' cut it out an' all. That was for Manning in Raglan Street. What lousy pay!

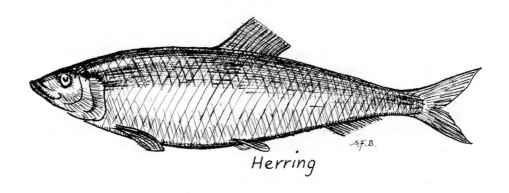

Herring

Branch, a Lowestoft lugger built to Cornish design in the 1880's. She has an early steam capstan amidships.

Post war Yarmouth drifter *Ocean Dawn,* YH77. The molgoggers are clearly seen on her bows.

CHAPTER THREE

Sail and Steam

'In Lowestoft a boat was laid,
Mark well what I do say!
And she was built for the herring trade,
But she has gone a-rovin', a-rovin', a-rovin'
The Lord knows where!'
  (Rudyard Kipling — 'The Lowestoft Boat')

Some of the fishermen who speak in this book began their careers in the final stages of the transition from sail to steam and it will be helpful to first look briefly at the development of the ships they sailed.

The first boats that have to be mentioned are the three-masted luggers, which were particularly associated with Great Yarmouth and which incorporated both speed and beauty in their design. They were called luggers because all 3 masts (fore, main and mizzen) carried lugsails, but they did vary considerably in dimension. Their clinker-built hulls (i.e. with the planks overlapping) were anything from about 30 to 60 feet in length, with a gross tonnage of somewhere between 38 and 80 tons, while the number of crew they carried usually ranged from 8 to 11 men, depending on size. About the middle of the century came an important modification in their design, when the mainmast was done away with. This gave more working room amidships and turned them into two-masted luggers; and it is these boats that figure so prominently in many contemporary seascapes. One specific and interesting feature of the lugger's crewing arrangements was the hiring of men from the inland country districts just to power the hand-capstan when hauling the nets. This was the origin of the joskins or half-breed fishermen, who left the land after the harvest was over and then did an autumn season herring-catching. It was their muscle that pulled in anything up to 2 miles of heavy, sodden warp, together with the nets attached.

During the 1870's came a further development in the design of the East Anglian fishing boats — the abandonment of the lugsail in favour of the dandy or ketch rig. This divided the total sail area, so instead of having just two large, square sails there was now a greater number of smaller ones. Both the mainsail and mizzen sail had gaffs; there were topsails on both masts; and a foresail was often worked on the foremast (which now became known as the mainmast) in addition to the mainsail. The use of the bowsprit with its jibs was continued from the luggers, but the outrigger and standing lugsail aft disappeared. This is a very simplified account of the changes in sail plan and, it is to be hoped, not too confusing in its brevity. Just to add to the confusion, the fishermen themselves still continued to refer to the new dandies as luggers — even though their appearance was considerably altered and even though many of the boats were now carvel-built instead of clinker.

There is more than óne possible explanation as to why the sailing drifters changed from lugsail to dandy rig, so it is wise therefore not to be too dogmatic in ascribing causes. One possible reason is that many local boats went trawling in the winter months after the herring season was over, but the lugsail rig wasn't very suitable for the trawling technique. When the Hewett fleet of smacks moved up to Gorleston from Barking in the 1850's, many of the East Anglian vessels began to see the advantages of another sail plan — and subsequently adopted it. An added advantage of the dandy rig over the lugsail was that it gave the boats greater manoeuvrability and ease of handling in the tricky North Sea shoals and shallows, no small consideration in the days when owners were not always too fussy about whom they crewed their boats with. In fact, it was not at all unknown for men of no previous experience to be taken on, or to find themselves heading out on a North Sea trip as a result of the 'Shanghai treatment'.

These boats must have constituted a marvellous spectacle in the last two decades of the 19th century. Their cleanness of line was a tribute to the designers, their durability and performance a testimony to the skill that went into their construction. A number of local yards made a name for themselves by the quality of their work, among the most notable being W. Teasdel, J.H. Fellows and James Critten of Yarmouth and Samuel Richards, Henry Reynolds and Fuller & Chambers of Lowestoft. Even William Brighton of Surlingham, a man more renowned for the building of Norfolk wherries at his Broadland yard, produced sailing drifters. Inevitably, there was a great deal of pride felt by the crews regarding their own particular boats and this in turn led to rivalry. Where speed was concerned, it was generally admitted that *Paradox (YH 951)*, owned by the Haylett family of Caister, was the fastest of them all. "Went like a racehorse," it was said of her. There is a huge, blown-up photograph of this drifter that occupies practically the whole of one wall in 'The Ship' public house, in the old part of Caister village. It is well worth going in to see. What it must have been to have seen the real thing, in great numbers, as they headed out to the fishing grounds under full sail (the great canvas cloths dressed with a mixture of horse fat, salt water and red or yellow ochre), or as they came speeding home, straining every spar and stay to make a good price at the market.

Steam began to compete on the fishing grounds in the early 1880's but was a long time coming to the Suffolk port. Then in 1897, the Diamond Jubilee year of Queen Victoria, Lowestoft saw two notable events: the opening of the swing-bridge across the harbour and the launching (in Chambers & Colby's shipyard) of what is usually acknowledged to be the first East Anglian built steam drifter. Her name was *Consolation (LT 718)* and she was built for George Catchpole of Kessingland. Other local owners followed his lead when they saw how successful steam power was, and from then on local shipyards began to produce steam drifters at an impressive rate. There were around 250 sailing drifters in Lowestoft in 1898 and only one steamer; by 1914 there were about 350 steamers and only a handful of sailing boats. It was a similar story in Yarmouth.

The economics of the change-over are not without interest either. In 1900, the last year practically that sailing drifters were built in Lowestoft, the total cost of a finished vessel was around £600 for a boat of some 60 feet in length and 38 tons net. By comparison the steam drifter cost nearly three times as much, but it was nearly 20 feet longer and contained a good deal of expensive machinery. Its advantages over the sailing boat were obvious to everyone, however: it wasn't so directly at the mercy of the weather; it was faster on the journeys to and from the fishing grounds; there were no regular charges incurred for towage by one of the G.E.R. company tugs out of harbour or back in again at the end of a

SAIL AND STEAM

Dutch "Hoy" C. 1790

Dutch, Katwijk beach, sailing drifter. Circa 1900

trip (the sailing boats often had to be towed in or out because of head winds or calms); and there was greater storage space on board for nets and fish. All of these considerations played their part in persuading owners to go in for the larger capital outlay involved in purchasing and fitting out a steamboat. One noticeable thing about the first decade of the 20th century, incidentally, is the number of fishing syndicates and companies that were formed, in order to cope with a generally more expensive way of operating. The tendency in the days of sail had been for owners to have small fleets of boats, very much family concerns, and there were also a good number of individual owner-skippers in both Lowestoft and Yarmouth.

The earliest steam drifters were really nothing more than dandies with an engine fitted and with the necessary alterations to the stern post in order to take the propeller shaft. They all carried a full suit of sails, in case the engine failed, and there was no wheelhouse (some didn't even have a ship's wheel, in fact, but kept the tiller). The only visible evidence of machinery of any kind was the narrow Woodbine funnel which sprouted from its mounting in front of the mizzen mast and hoodway.

Within a year or two wheelhouses began to appear, often with the funnel placed in front. It was still the Woodbine variety, rising from a 15 HP engine, and it gave the boats a very distinctive look. By about 1903-4 funnels were almost entirely aft of the wheelhouse, an altogether more logical siting, and the classic lines of the steam drifter were now clearly established. They were beautiful boats, perfect for the job they had to do, combining centuries of hull design with modern technology. It was a happy marriage, and the years up to World War One saw an ever-increasing stream of top-class boats produced. For the most part they were constructed of wood, but there were steel ones as well, and both kinds were often completed in as little as two months. The fastest ever built was the *Briton (LT 1017)*, which had its keel laid in the famous yard belonging to Sam Richards (on the south side of Lowestoft's Lake Lothing), and which was ready for sea 40 days later. The year was 1906 and the total cost was £2,370. Such speed and efficiency must have given both builder and owner great satisfaction.

As far as dimensions are concerned, both wooden and steel steam drifters were somewhere in the following range: 75 feet long by 18 feet beam by 8 feet deep up to 85 feet long by 19½ feet beam by 9 feet deep, with a consequent net tonnage of anything from 23 to 50 tons. Some of the bigger drifter-trawlers went up to over 90 feet in length, for just as there were boats in the era of sail that did both kinds of fishing (converter smacks), so there were too in the days of steam. These vessels' versatility was very much in their favour, because whereas many of the drifters were laid up in between herring voyages the drifter-trawlers were able to keep working. Many of them are still talked about today for their money-earning capability, particularly in Lowestoft, which was the port for drifter-trawlers. Boats like *Hosanna (LT 167)*, *Merbreeze (LT 253)*, *Sarah Hide (LT 1157)*, *Margaret Hide (LT 746)* and *Byng (LT 632)* have become part of local folklore and will remain so while the men who remember them still live.

There is a romance too attaching to the engines that powered the boats. They were of the steam-reciprocating type, with coal-fired boilers, and many of them were made locally. Elliott & Garrood of Beccles were one of the leading firms in this field, manufacturing a variety of engines from the early, simple, two-cylinder pots (with the famous Woodbine

SAIL AND STEAM

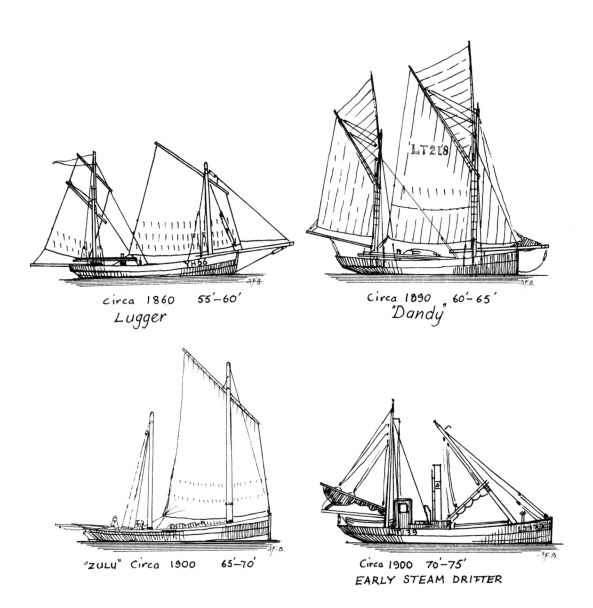

funnel) to the more complex and powerful monkey-triples and compounds. There were a whole range of triple expansion and compound surface condensing engines produced for the steam drifters, and not just by Elliott's. Samuel Richards & Co. Ltd. of Lowestoft, shipbuilders, made many of the engines used in their own boats, while Davis & Co. of Oulton Broad were also marine engineers. Over in Yarmouth the big names were Crabtree, Burrell and Pertwee & Back.

Apart from its importance in driving the drifters' engines, steam power was also necessary for working the capstans that hauled in the nets. It had been used for this purpose since the 1880's, particularly after Messrs. Elliott & Garrood had developed a new and vitally important piece of equipment that rendered all previous models obsolete and inefficient. The principle was basically simple; what they did was to have the steam pipe and the exhaust pipe pass up through a hollow steel tube in the middle of the capstan to the drive unit, which was mounted on top. The propulsion was then transferred to the outer barrel by a pinion that engaged a horizontal bevel wheel, and this caused the barrel to rotate. There was also a pulley wheel attached to the drive unit, which could be operated independently and which was used for unloading the catch. The Elliott patent capstan was capable of winding in the warp at a rate far beyond that possible by hand power and its use soon became universal. However, its introduction was stormy in the initial stages because owners customarily took a share of the boat's earnings in order to pay for the new equipment; and this led to the so-called Capstan Riots of the 1880's. It is interesting to note that even at the very end of the herring fishing, 70 years later, owners were still taking out a share on behalf of the capstan.

The engines called for new men on board. There were two of them, the chief engineer (who was always known as the driver) and the stoker, and they kept up these titles even when the motor boats came in. The first motor fishing boats were smacks or dandies that had been converted to oil propulsion. The first purpose built one in Lowestoft appeared in 1926, the drifter-trawler *Veracity (LT 311)*, with her 200 HP Deutz engine. It was still some years before the diesel engine superseded steam and became generally accepted as the desirable power unit for fishing boats. Rocketing coal prices during the 1930's had a good deal to do with the eventual change and in the immediate post-war period new classes of drifter-trawler began to emerge. They were very different boats from the ones that Jack Sturman (1891-1978) cut his teeth on:

'I went driver about 1912. I'd bin stoker afore that. Course, I lived at Mutford in them days. There was a lot o' good fishermen out there then, but now I don't know if there's any. Most of us biked down inta Low'stoft, used to leave our bikes at a shop there near the harbour. That was hot down in the engine-room, but o' course you had your skylights up, weather permittin'. I suppose you got used to it and you didn't think nothin' about it. The boats used t' burn about 10 or a dozen ton o' coal a week, and o' course the coal merchants used t' have their runners come round and go t' the skipper about the coal and that like. That didn't matter what sort o' coal that was; the skipper would go where he could git the most discount, see. You didn't have no say. You had t' burn it, but you didn't have no say in, it. The skipper had the privilege o' payin' the bills and that like, and gittin' the discount. The driver didn't have nothin' t' do with it. When you were herrin' catchin', like at Low'stoft, the little ol' coal boats would come alongside you there in the harbour and you

SAIL AND STEAM

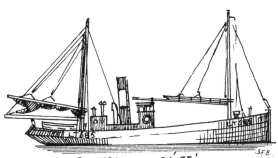

Circa 1910   80-85'
"CLASSIC" STEAM DRIFTER

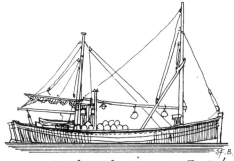

Converted "FIFIE" Circa 1925   70'

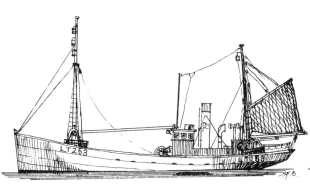

Coal burning Drifter-Trawler
- built Lowestoft 1930

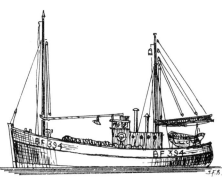

SCOTTISH, PURPOSE BUILT
MOTOR DRIFTER. 1930's — onwards.

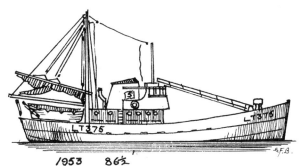

1953   86½'
POST WAR DIESEL DRIFTER-TRAWLER

were always full up. But at Shetland, if you were workin' Shetland, they'd bring it down on the cart and coal you time you were layin' there. You could hold about 10 or a dozen ton, and o' course the drifters what went trawlin' as well, they had a bunker in the after-end o' the hold — for fore-side o' the wheelhouse, you see. Course, you didn't stop at sea 12 days then. If you were gone a week, that was as much as you did. Gettin' towards the end o' the trip, you'd say t' the skipper, "There's only so much coal left." And he'd know that you'd gotta go in.

'When I first went driver afore the First World War I went in a drifter they called "The Three Ducks". Her real name was the *Waveney*, but her number was *LT 222* and that's why they called her "The Three Ducks". I was there a year, driver, and then I thought I'd git somewhere else. I was told about a drifter, the *Homeland (LT 125)*, and so I went and see the owner, Mr. Jenner, and I got the berth there. That'd be about the beginnin' o' 1914. Well, I went t' Shetland in the *Homeland* that year and we got orders t' proceed hoom because the war had started, you see, and we weren't allowed t' fish. And o' course we come hoom and the ship was laid up and I went t' work with my father down at Youngman's Brewery. Then they wanted me t' go driver agin in the *Homeland* because the government had took her over for minesweepin' or patrollin'. And we had t' go on the South Pier for a medical. On Low'stoft Pier, that was where I went up to where the doctor's office was. I knocked on the door and he sung out, "Come in." When I got in he looked at me. He said, "What department?" I said, "Engine room." He said, "You'll do." And that was the medical examination I had! And o' course we went and got the ship ready. She was laid up at Southwold, up the river there. There was a lot o' drifters laid up there, you see, after they got stopped when the First War started. Anyhow, we went and got the drifter from there, brought her t' Low'stoft and fitted her with a crew. The second (engineer) I had was a bricklayer! And some o' the others were farm labourers; they come from Thorpeness and Aldeburgh, places like that.

'That drifter I went in, the *Waveney*, what I was tellin' you about, she had the wheelhouse the back side o' the funnel. The wheelhouse used t' be over the hatchway where you went down t' the cabin. That was a Chambers boat; so was the *Forethought (LT 27)* what I was in later. Chambers's shipyard was up against where the railway sleeper dump was, and then there was yards on the other side o' the water as well. Then, o' course, there were shipbuilders at Yarmouth. Fellows was one lot. Elliott & Garrood at Beccles made engines. And Crabtree's at Yarmouth, they built engines too. The Scotchmen used t' call the Elliotts "Beekles engines" — "Have you got a Beekles in?" they'd say. Now some o' them Elliotts were very fast. They had the high-pressure cylinder on top o' the intermediate cylinder and the low-pressure cylinder. You see, most triples had the three cylinders out-length, but the Elliott monkey triples had this one cylinder on top o' the others. That way they didn't take up so much room, but they were higher o' course.

'The drifters used t' make about 8 or 9 knots when they were steamin'. Some were faster than others, but about 8 knots was a decent speed. One ship I was in, *Maybird (LT 444)* she was very fast. They used t' say she didn't care for no one. We were round there at West'ard once, round Padstow, and the chap what was second t' me, he say, "There's so and so poppin' up alongside." I say, "All right, let him go." Well, when we come in, this here skipper he say t' me, "You could have catched us if you'da liked, I bet." I say, "Yes, but we didn't like!" Oh, she was fast, that boat. She was a steel boat, and when we were runnin' the trial on her I set down in the engine room along o' the engineers from Chambers's, what'd built her. After a while I say t' the foreman engineer, "That rod is a-gettin' hot."

He say, "Oh, how do you know that?"

I say, "Look, can't you smell it?"

And so he say, "Well, what's that t' do with you?"

I say, "Well, when you've done with her, I'm goin' to have charge of her, so that's all that is t' do with me!" And o' course that was right. You see, the connectin' rods sometimes got dirty. But the funny thing was if you swabbed 'em with parafeen, that cooled 'em down. I knew a chap what'd been a driver a long time and he told me that. "Do you know," he said, "if you swab it with parafeen," he said, "that'll take the muck off the rod and cool it down." The muck caused friction in the packing, see, and the parafeen cleaned it orf.

'I've started from Low'stoft and bin right round the British Isles. I've even been t' Stornoway in the Hebrides, driftin'. You'd be there at the beginnin' o' May and then work round t' Shetland. That used t' be hot down below in the summer. You didn't have anything on, only your dungarees. Freddie Mills from Corton, he used t' wear a swimsuit; that's all he'd have on in that engine room. When you were drivin' you done four hours or so and then your stoker had a spell. That was if you were on a long steam. If you were workin' out o' Low'stoft, fishin' time, you'd drive her in when you'd done haulin'. Then o' course you'd land and be away agin, so the stoker would have a spell goin' orf. Praps you'd say, "All right, let me have an hour or so and then I'll take her." O' course, time they're landin', you're a-cleanin' your fires out and gittin' ready for goin' out agin. See, if they're a-puttin' some coal in, you're a-slicin' it down and puttin' your bunker lids on and that. If you git time!

'If you blew down and then got up steam agin, that'd take you about 10 or 12 hours. That all depended. You didn't rush it, you see. And o' course you'd got a good lot o' water there. I couldn't tell yuh how much exactly, but there was a good lot. And o' course you carried 160, perhaps 180, pound o' steam pressure. When they build the *Forethought*, they put an engine in her out of another boat. That was a Beccles engine, so that went over t' Beccles t' be done up and then that was put in the *Forethought*. When we come t' run the trial we didn't git the speed out of her that we thought we would. But o' course we went orf t' Shetland in her and when we got in the harbour, afore we'd hardly hit the quay, she went up "Sssshhhh". That blew orf, you see. Yes, she blew orf for a few minutes and then shut down. Then after a few more minutes up she go agin. That turned out that the Elliott's people, unbeknown to anybody, had lined the high pressure cylinder and made it smaller so she wouldn't burn so much coal. They were experimentin'. Nobody knew anything about it. We only found out when we got hoom. We couldn't git no more'n seven mile an hour out of her when she was doin' her best.

'When that was a new ship and you went on your trial, you had a whole shipload. You'd take your wives and that like, you see. You'd go out for a steam; you know, out in the roads. But all that got stopped when that come in so everybody had to have lifebelts. See, you only praps carried a couple or three lifebelts and you had 30 or 40 people on board. Ol' man Jenkins would be there on the South Pier, a-takin' photos and that like. Course, they arrested him in the First World War for bein' a German spy! Just because he was takin' photos. And, I mean, he lived in Low'stoft and everybody knew him! (H. Jenkins was a well known local photographer).

'When you were driver you done certain jobs on the engines; but the things you wanted done and couldn't do yourself, you had a fitter come. I remember Billy Hid (William Head, one of the leading Lowestoft vessel owners) sayin' t' me one day, "Chief," he say, "they aren't no good t' me if they can't keep goin'." But normally Elliott's engines would last a lifetime. Well, that isn't so long since there was some of 'em kickin' about somewhere. I wouldn't know where you'd find one now, but they were good. They had white metal bearings. A lot of the others used t' be brass, and if you didn't watch 'em the brass would heat up. Course, after you'd been driver a little while, you'd soon know whether there was somewhere gittin' warm by the look of it. That'd be dry; there wouldn't be any oil showin'. Course, they weren't like a motor; they weren't in anything cased in. You could see all the works. You had oil cups on the front o' your engine t' oil some parts, and then you filled your bearings up about every 4 hours because you'd got cups on them as well. You'd have feeders t' do the lubrication with and you'd count so many drops of oil for each part. You used just the one oil for the lot, though. You got through praps 40 or 50 gallons in a voyage. Some of the ol' drivers, they used t' try and run special with it (you know, not use more than one another in a week) and you'd hear 'em say, "Oh well, we didn't only use so much."

'I had a breakdown once at Shetland. We were comin' in and we broke a bearing. So o' course I stopped the engine and I say, "I've got a breakdown." The skipper say, "Well, what are you goin' t' do about it?" I say, "Well, I don't know yit." Anyway, some o' the crew come scrabbin' down the engine room, so I say, "Clear you orf!" I say. "Go and fetch me that foc'sle funnel. Go you and do that, and that'll help." I had to turn up the induction pipe what went into the condenser. We made like a traction engine, so she exhausted out into the sky through this here foc'sle funnel. That worked all right till we got in, you see. We didn't have much sail, only the mizzen. Praps you had a tow-foresail, but then the wind had gotta be right for that. In the early boats they did have a mainsail at first as well, but in the latter part o' the time they done away with that. The foremast was only there for a derrick when you landed. Nowadays, when they break down, they tow 'em in; but in the old days you got 'em in if you could.'

Horace Thrower (born 1904) spent nearly the whole of his working life at sea as a stoker or second engineer on fishing boats and still remembers the details of his steam engines:

'The first engine I looked after was a Richards. That was in the *Nil Desperandum (LT 175)*. Then I went on to an Elliott's triple in the *Empire's Heroes (LT 703)*. An Elliott's triple she was. They were good engines; you could rely on them. They'd got a big flywheel, see, and that helped the engine round. That took a lot o' wear and tear orf the engine. Yes, they were damn good engines, they were, and they were very economical too when you were burnin' the coal. Yes, they were very light on coal. You could use an ordinary dustpan t' coal up on them sometimes. You could! You know, you could just sprinkle a little on the furnace. They were a lovely job, they were. Then you got the bigger ones, like the Burrell's out-lengths, what they called an out-length triple. They were hungry, they were; they'd chuff some coal up! The Elliott triples had one cylinder on top of the other two and you could handle them lovely. They were very easy t' handle and very economical. Very little noise too, even with that big flywheel. You could git about 11 mile an hour out of 'em when you were steamin', and that was a good speed at that time o' day.

'Them Elliott's triples would use about 15 ton o' coal a week, I spose, on the hoom fishin'. And sometimes praps you didn't even want that. See, you never got low; you'd always keep fillin' up as you come in. The ol' wherries used t' come around alongside of you and put it aboard. That was all loose at first and then they got bags later on. Sometimes if you couldn't get it, you just didn't bother; you went t' sea and waited till the next day. That all depended on how many boats was in, you see. If there was only a few boats in and you got the chance t' coal, then you'd coal. That was just the same with water. You had a freshwater tank for drinkin' and then another one for the boilers.

'The cook used t' use sea water for the vegetables, and you used t' put it in the boilers as well. Yes, you'd put that in sometimes. But you wouldn't put it in if you could help it because that would fur all the boiler up, see; all scale the boiler. You'd gotta clean the boiler out praps every six weeks if you drew sea water. You could soon tell when you got foul; you know, when the tubes got fouled up. That was harder work steamin'. See, that's why you cleaned 'em out every so often. That was only a matter of 24 hours and you were at sea agin. You'd just blow down and let 'em stand for a few hours till they got cool and then fill 'em up fresh. You'd let your steam blow right out and then you'd go right inside the boiler and clean it all out. You used t' have a chipper and just scrape 'em more or less. That'd soon flake orf. Yes, that wouldn't be long flakin' orf once that got cold. You used t' clean the pipes just the same, just poke somethin' through 'em. That was like a big paint scraper, a long thing with a blade. That didn't take long.

'When you were herrin' catchin' you didn't use t' clean the furnace out every day. You'd just rough it over. But when you were trawlin' you had t' practically clean the fires nearly out. You had two, you see, so you'd clean one out, light it up fresh and then go on the other one. Course, your steam would go back a bit, but you had t' do it otherwise you'd have clogged up. You'd lift the clinker up orf the bars and that would be like one big cake. When they got like that, you couldn't steam because you weren't getting the draught, you see. You'd be surprised at the muck what come out of a furnace. You know, you'd rake the back ends, where all the soot and that go. That all lay in there and burn, so you had t' git out. When you were herrin' catchin', you'd rake the fire out after you'd shot the nets. You'd let the steam go back, see, so that weren't too hot and then you'd clean 'em right out. You'd be finished in about half an hour. You used t' chuck the cinders overboard; put it in buckets, pull it up the ventilator, stand on the deck and chuck it over. You know the two ventilators? Well, they've got a pulley up the top and there's a door there as well, where you pull the buckets to and git the clinker out.

'That used t' take about 12 hours t' git a proper hid o' steam when you were firin' orf. Especially when they were cold. Yes, that'd be a good 12 hours then t' git 'em goin' properly. When you come in for the night on herrin' catchin', if you got time, you'd just rough the fire out. Just git some of the worst of it out, the worst clinker out. You'd push the fire back on one side, clean out where that'd been, git your slice and bring it back over, then clean out the other side of the furnace. You'd got a fire there all the time, you see; all you were doin' was movin' it across while you cleaned out. Once you'd done, you oughta hear the hummin' noise! You know, when you'd cleaned the fires out, they'd hum like the devil. Yes. That was the draught. See, you'd pull the draught places away at the bottom, git 'em about halfway open and then you'd hear her start t' roar. When you didn't want no steam, like when you're lettin' her die down, you'd shut the dampers up.

'On the Elliott triples you had two fires. That was a double furnace. Course, the little Elliott pots on them early drifters, they were a single fire. And they had an upright boiler. The ones what I worked with had a long boiler, but the pots were an upright one. There weren't much power in them. I don't know what horsepower they were, but that'd be about 15 I should say. The triples were more powerful, though, and o' course Elliott's triples would do more revs than what an ordinary out-length triple would. Oh yes, they were higher revvin'. That was that big flywheel, see, helpin' the engine go round. That was surprisin' what a difference that made. They weren't much trouble either, them engines. That they weren't. You could do a lot of the jobs on 'em yourself, like packing the pistons and that sort o' thing. You used what they called greasy packing for that job. You'd take out all the old packing what was burnt and put all new greasy packing back. That was cottony stuff and you'd cut what length you wanted and push it up. Then you'd put the gland back on and tighten up.

'You had little oil feeders for lubricating the engine. You had t' hold them in your hand, see, and use them when she was a-runnin'. You did it when you thought of it; there was no set time for it. You'd be sitting down and you'd say to yourself, "I'd better oil her up in 5 minutes." See, you'd got the cups and they were always full. You kept them up. They were on the main bearings and you kept them full. The other parts, they'd just got holes and you'd put the feeder in and oil round. You had very little trouble with the bearings on them engines. When you stopped once a year and had a refit, you'd have all them out. You'd take all the bearings and everything out and have a look at them. Praps you'd have t' file out where the oil run through the bearing. Course, that was only white metal. You had a proper thing and you'd just bore a hole and make a runway for the oil, so that when that was closed up that would always be full of oil.

'The ol' skippers, they always wanted more than an engine could give. And I tell you what I've done, though praps I shouldn't say so — fill the ash buckets up with wet ashes and put them on top o' the boilers, on top o' the valves. That'd keep the valves down so they wouldn't blow orf; give you a little more steam. You'd do that if you'd got a good catch o' herrin' and you wanted t' git t' the market. Oh yes, we've done that more than once! Course, you'd have got wrong if the Board o' Trade ud found out. You weren't sposed t' do that. See, the boilers were set for a certain amount o' steam and you'd gotta keep to it. When you got above a certain pressure, the valves would lift and blow orf. See, when they got to a certain pressure, they started. When you put the ash bucket on top, you had t' keep a watch on your steam. You mustn't let it git too high! They used t' blow at about 180 pounds a square inch so you'd know how far you could go. Normally, you used t' keep at about 170/175. Course, there was times when you praps stopped for somethin' and you'd got big fires — well, that would build up and blow then. They were safety valves, see, and praps they'd blow 5 or 10 minutes afore they dropped agin. That was surprisin' the steam you lost that way. If you'd had 180 pressure, when you looked agin praps you'd only got 160. She'd lost about 20 pound!

'When you tanned your nets on board, you had a little valve for that foreside o' the engine room casing, up on deck. There was another valve down below what controlled the water, so you'd turn both them on when you wanted t' tan. You'd fix the pipe on up on deck and then you'd be away. You used t' have a long copper pipe for tanning, and you could bend it all ways so you could git it into the tank. You'd run that down inta the bottom o' the tank so you didn't git scalt. If you only had it halfway down, that was liable t' fly out! See, you'd got full pressure there when you tanned. You'd turn it on down in the engine room and then there was a safety valve on the pipe itself, so the man who was up there on deck could ease it down if he wanted to. He'd just turn it down t' what he wanted.

'You had a separate valve run orf the main engine for the capstan as well. The capstans originated from Elliott's. They were the ones what made all the capstans, Elliott's. They were a good ol' thing and all; they used t' do some work! I mean, they used t' haul in all the ropes and all the nets. Well, you used them for practically everything. When you landed, you used them t' land your herrin'; when you were mooring up and you couldn't git in, you'd pull the boat up on the capstan. There you are — capstan again! You had a steam pipe in the centre o' them, that ran up inside, and if anything went wrong you could git at it easy. If the pulley went wrong, or they got warm or anything, you could soon take 'em apart and put a new liner or somethin' in. Down below you had a box on the side o' the engine where you had all the different valves. You know, the one for the capstan, one for tanning with, one for the auxiliary engine, all that sort o' thing. You had a little steam engine, an auxiliary, t' run the lights, and that had a valve because that ran orf the main boiler as well. Yes, all the different valves were in this box.

'You used t' carry quite a few spare parts with you on a boat — bearings and that sort o' thing. Oh yes. And packings and that sort o' stuff. Course, you never knew when you might want them sort o' things. See, when the engine was runnin', you'd oil the pistons t' keep the packing moist. But if you let 'em go too long, they'd start t' burn and pieces would keep fallin' out and you'd lose steam. Mind you, that was very rare you let 'em do that. If you see one leaking a bit, you'd just tighten up the gland what push the packing up. That'd got two bolts at each end and you just screwed them up a little bit tighter, so that would stop it. When the gland got right up the top o' the cylinder, then you knew what you'd gotta do. That'd all gotta be re-packed then. That weren't much of a job. We used t' have 'em all ready, so if we wanted one in a hurry we'd just fish it out and put it in. If that was anything too tricky, though, you'd have the engineers come and do that. You wouldn't do it yourself, no. See, like when you had a refit, you'd have the hids (heads) orf and everything out. All new rings and everything were put in. They'd do that generally once a year — you know, strip 'em right down. The boat would go into the dry dock for that. Course, you'd be workin' aboard yourself. You know, helpin' like and doin' the odd jobs. But the main part o' the engine, the engineers would do that, the fitters.

'Another thing what come from the engine room was the deck-hose. You used t' work that orf the auxiliary engine. And that worked the lights, that auxiliary. When I first went the lights were carbide, but then they gradually converted them so we had all electric. You had a little dynamo put in then; that was driv by steam orf the auxiliary. Carbide lamps were a nuisance, but I weren't with them much, only when I first started. On the trawlers there was a man t' see after all that, the lights and everything. Yes, you used t' carry a trimmer. He'd trim all the coal and the lights, pull the ashes up, all that kind o' thing. But on the drifters that was the stoker's and the engineer's job. Oh yes, the ol' driver had t' be in as well; he had t' do his share. If that was fine when you were shootin', he used t' clean the fires then. Course, you don't want much steam when you're shootin' your nets, just your engine tickin' over that's all. There were plenty o' times when you were shootin' when you just used the mizzen. And even then the engine would still go round, you know. As you were goin' through the water the propeller would keep turnin' and the engine would turn as well. Yes, even when you'd shut orf, that'd do that.

'There weren't much danger on the engine really. I mean, you'd got a guard rail in front of you when you went round, so all you'd gotta do was git hold o' that when that was bad weather. Course, there were bad accidents known because you could slip in. See, there was only that one rail round. That was about breast high, so you could slip underneath it if you weren't careful. And that has been known for people t' fall in that way, you know. Oh yes. There was a boat steamin' t' Ymuiden one time with a catch of herrin' and they missed the second engineer, the stoker. They couldn't make out where he'd got to. And they went down below and there he was, all mangled up in the engine, goin' round the bottom in the dill. They couldn't recognise him. He was simply mangled up t' nothin'. He was in where all the oil and that is, see. That was all hollow under the engine; you could crawl along underneath there. Yes, that's where they found him. How he got there I don't know. No one know how it really happened. See, the skipper said he found the engine kept easin' up, easin' up, so away he come aft and he couldn't find the second engineer nowhere. So he went and called the chief out, and when they went and looked there he was — down in the dill! That was a long, long while ago. I can't remember the man's name and I forget the name o' the boat.

'Another thing about the engine room is that it used t' git hot down there when you were afore the wind. You know, when you was runnin' afore the wind. Oh, that'd be devilish hot then! You couldn't hardly put your hand on the handrail sometimes. The heat couldn't git away, you see; that was all blowin' forward. When you were steamin' hid t' wind or broadside that would blow it away, but when you're afore the wind you'd be surprised how hot that git. I used t' turn the ventilators on so that could git away, but even that didn't always work. You noticed the difference between summer and winter as well. Sometimes in the winter, when that was right cold, you had t' shut all the blinkin' doors. Then summertime that'd be just the blinkin' opposite — you wanted all the doors open! You couldn't win on that. I mean, in the winter, after you'd fired up and oiled round, you'd sit down and have a read for five minutes and you'd be surprised how cold that was. Course, your doors were right at the top o' the steps and you'd be sittin' at the bottom, so you got all the draught. I tell you, I've seen the time when you had t' shut them doors t' keep warm. And then summertime that'd be just the blinkin' opposite! You'd want everywhere open. Course, as far as the boats themselves went, a steamboat was a lot warmer than a motor one. Blimey yes, a lot warmer! Well, I mean, they were warm over practically all the ship. They'd got that bulkhid, see, between the engine room and the cabin, and that was always warm so the cabin was warm as well. A wooden boat was warmer than a steel one too. Oh yes.

'The first boat I went in, the *Nil Desperandum*, she was a wooden boat. She was the only wooden one I went in. They were lovely ol' sea ships and all. That they were. A darn sight better than steel boats. You'd be surprised at the weather they used t' stand, them there wooden boats. They used t' ride the sea just like a duck. That they did! They were just the same when you were steamin' as well. You'd be goin' along there with your mizzen up, steady as a rock. Or if there was a nice breeze sometimes, you'd have the big sail up too. You'd go along with that and you wouldn't know you were at sea hardly. Oh yes, you used a tow-foresail. That was a huge thing, shaped like a big ol' triangle, and you'd be surprised the wind that'd hold. You'd pull the mast up orf the top o' the wheelhouse, (on steamers it was always lowered to rest on the wheelhouse when not in use) run the sail up, bring it aft and make it fast. That used t' go right aft, nearly t' the blinkin' galley. You'd make it fast on the rail somewhere and that was surprisin' what a difference in speed that made to you. That'd make about a couple o' mile an hour extra time you were steamin'. And you were steady with it too. Yes, you'd use that a lot when you'd got a nice breeze, specially if you'd got any distance t' go. Praps you'd got 50 or 60 mile t' steam — well, that'd pay you t' pull it up. You'd git there a lot quicker; praps save yourself an hour or so.'

Ned Mullender (born 1896), a man whose long career at sea saw him do nearly every kind of job on every kind of fishing vessel, also remembers the early engines and what life was like below deck:

'I remember when I was in the *Excelsior (LT 698)*, she had a little ol' Elliott pot engine with a stand-up boiler. They were all right, them engines. Yeah, they were all right. We were comin' hoom once — I was single then — and we'd bin lining (long lining for cod). The skipper, he come t' me and say, "Neddy!"

I say, "What?"

He say, "If you don't stir this boat up, you won't see your young woman tonight."

I say, "Oh, that's like that, is it?"

He say, "Yis, that is."

Well, I went down below and I say t' the chief, "All right, chief, I'll give you a spell."

He say, "Oh, I don't want no spell."

I say, "Yis, you do. Go on, I'll look after her."

And after he was out o' the engine room I turned the whiffler on. They had what was called a whiffler t' make a draught t' draw the fire up. Well, he never had that on; he was an oldish man and he was always scared the boiler was goin' t' bust. Anyway, I turned it on and that weren't long afore I had a full head o' steam. No, that weren't long afore the ol' engine was dancin'. The thrush was goin' like billy-o. That was quite laughable really, but we got hoom all right.

'The horse power wasn't much on them little pots. Them upright boilers only operated at a pressure of about 120 to 130. They had a plug on them, you know, made o' lead. You know, you burnt that plug so that let the steam out when the boiler run dry. You had t' keep that boiler pretty full. That'd want regular tending to. You couldn't go t' sleep on the job. Oh no, you'd gotta have a full pressure o' steam t' keep them running at any speed at all. But when they built the *Confier (LT 658)* and *Mary Adeline (LT 650)* and two or three more o' that class, they had a bigger engine with a lay-down boiler. But they still only had a single furnace. The *Heather Bloom (LT 1112)* had an Elliott's compound in her. Then there was the monkey triples. They had that cylinder on top to' the other 2 and they swung a big flywheel. I spect that flywheel weighed nearly ¾ of a ton. I should say so. A good weight anyway. Course, I might have exaggerated a bit on the weight of 'em because I aren't absolutely sure, but they were very heavy. The *Faithful Friend (LT 33)* had a monkey triple in her and she was fast. Mind you, there was two classes o' them engines. There was a smaller one and a bigger one — Elliott's 60's and Elliott's 80's.

'Burrell of Yarmouth made good engines as well. The old *Pearl (LT 461)* had a Burrell in her. They made 'em at Yarmouth; they had a firm there. A lot o' the Yarmouth boats had Burrell engines in 'em. So did the ol' *Pearl*, like I say. That had a slide-valve aft and for'ad. Yes, instead of just the round valve for'ad that had a slide one. Well that used t' hang up at times, git stuck. And if you got that locked, you couldn't git your reversing lever one way or the other. If you didn't watch out, that would stick. You'd have steamlock then, and the only thing you could do was t' open the drain-cocks and let it drop. Once you'd got them open, the ol' hand on the dial would fall back.

'The Elliott engines were very economical. You used t' have the coal boats come alongside, or you went alongside of a coal wherry, and they used t' put in what you wanted. If the chief said about seven or eight ton, that was what they used t' put in you. The ol' chief sometimes got a backhander o' course. I know one who got a good backhander once and he got the sack afterwards! When we got hoom at the end o' the voyage we'd burnt twice as much coal as anybody else — the same class o' ships! But o' course we hadn't really. He only said he'd took the extra coal on, see, and then took a backhander orf the merchant for saying so. Course, the skippers used t' git a discount on a ton o' coal. You used t' usually git about two or three bob a ton. That was a skipper's perks — years ago I'm talking about when I was skipper o' the *Impregnable (LT 1118)*.

'When I was in her my guvnor had shares in Bessey & Palmer's, see, so I was duty bound t' deal orf Bessey & Palmer's. Well, about three-parts through one trawlin' voyage I was a-leanin' on some railings at Padstow dock there with a skipper, a-talkin' with another skipper by the name of Arthur Artis. We were talking to him and another skipper come up, and I'd got my wallet in my hip-pocket. Well, this other skipper patted on my hip-pocket and he say, "I bet that's somethin' full."

I say, "What do you mean?" I say, "If we go in that pub together," I say, "and there's a couple more skippers in there and I pay a round," I say, "there wouldn't be none left."

"Go on," he say. "You wanta tell the marines that!"

So I say, "What do you mean?"

"Well," he say, "if you don't know, I aren't a-goin' t' tell you." And nothin' more was said that morning.

Later on I said t' Arthur Artis, "Arthur, you're an older man than I am. What was he jumpin' at?"

He say, "Don't you know?"

I say, "No, I don't. I git my two shillins a ton and that's that. If he's a-thinkin' I git more—"

"Well," say Arthur, "Bessey & Palmer's have offered me 3/6d, and I spect he's got extra as well, even though he deal with Craske's."

So I went along t' Bessey & Palmer's man and said, "What'll you give me for the new lot o' coal?"

"A shillin' a ton," he say.

Well, that was a shillin' less than what I'd bin getting, so I went along to Bray & Parkins, a Padstow firm, and asked what they'd give me. They said 3/6d, so I took it. A few days later

I got a letter from my firm, asking why I'd changed. So I told 'em. In the end that was the Bessey & Palmer's bloke what got inta trouble for offering me less than he should have done!

'Now as regards accommodation, when I was in the *Scadaun (LT 1183)* there was only about four or five of us slept aft. All the rest were for'ad in the foc'sle. They had a cookin' stove and a lavatory down there as well. See, ol' Lord Dunraven had that built for his yacht — you know, for towing it about. Mind you, she fished as well. A gentleman by the name o' Turkey Goldspink was skipper of her and she had a Low'stoft crew. The cookin' stove down a foc'sle wasn't usual, but most o' the drifters had accommodation for'ad at one time o' day. Yes, they used t' sleep about four or five down there. The foc'sle disappeared as the ships got bigger and better, but some o' the chaps preferred t' sleep for'ad because there was always a stove in there t' keep it warm. In the *Impregnable* we had 10 bunks aft and a foc'sle, but no one slept down there. That was used for a storeroom for lamps and parafeen and carbide, that sort o' stuff. All the nets and that were in the fore-hold, which was another hold the fore side o' your fish-room.

'The boats' fish-holds were divided into pounds. Some had more'n others. Some had 5 in the main hold and then they'd have an after-pound with boards across. That was what they called an after-well sometimes, and that had bunker lids to it and stretched right across the ship. You know, breadthwise. Now the other pounds had bunker lids to 'em as well and that's what your herrin' ran down through when you were shaking out the nets. You kept your buffs and that in the wings, but your fish would go down into the hold. Not all the under-deck spaces had bunker lids to them, though. Oh no. See, you'd always got more compartments than bunker lids. I think when I was in the *Impregnable* she had two into the fore-hold, t' put salt down if you wanted to, and four each side in the kids. We had shutes aboard as well t' run the herrin' from pound to pound, but as a rule you took a board out so they used t' run in that way. When you wanted t' bring up the boat on an even keel, you'd take a board or two out on the opposite side so the fish would run across and level the boat up. You only used the shutes when you wanted t' stow the herrin' away. Say if you were hauling Saturday or Sunday morning and you didn't want to come in, well, you'd use the shutes then so you could fill all your spare wings up and leave enough space for the following morning, when you shot agin.

'The holds were built so that you could take the boards out because you had t' wash out regular. As soon as you finished gittin' your herrin' out, all your pound boards were shipped out and washed. So was the hold. Afore you put the nets down agin for the next trip you'd hose everything down. You runned the hose from the engine room; you had a proper donkey (engine) what would drive that. That was salt water what you were drawing up. You had your valves down in the engine room for doing all that sort o' thing. See, there was a case down there with the valves in — one for fresh water to the boiler from the tank, and one for sea water to the boiler. Some of the ol' boys used t' like the boiler t' be salted up a bit. Not salted up so the density was all wrong, but just salted up a bit. That wouldn't take no harm; you used to scale it out every so often. And so you had this chest of valves, and one was for the hose. There was a rare force come through there. You could hit a bloke in the face and knock him down if that was on full. You had scrubs too for cleaning down with; you know, brooms. You had to be very clean where fish was concerned.

'The water used t' run into the bilges after you'd done cleaning. You had a grating aft with a lot o' little holes in — say a little bit smaller than a threepenny bit — and all the waste stuff used to go there. And what stuff went through, like little bits o' head and scales, that accumulated in what they called a dead well. That was deeper and right aft in the hold. You used t' take this grating up and clean it out, so that was nice and clean, see, when they'd pumped all the water out. You could git your hand in there and git all the filth out. Well, all that sort o' thing was half o' keeping the ship in good trim. You had to keep your holds clean, and they used to be lovely and clean — specially if you had a driver what didn't mind putting a little power on the ol' hose! you know, give it the full valve. You used to wash down right for'ad and right aft as well, up on deck. Oh yes. I mean, if you didn't keep your ship clean — well, that was a poor do.'

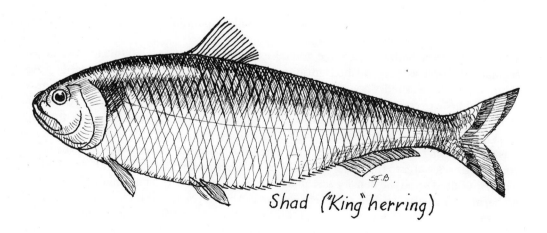

Shad ("King" herring)

Above, typically crowded scene in Lowestoft herring dock in the mid 1930's. The drifters are having to land their catches stem on because of the numbers wanting to get to the quay to unload. In the foreground *Lord Anson*, LT 344, *Go Ahead*, LT 534 and beyond her *Sunbeam*. Below, herring being scutched into quarter cran baskets for hoisting ashore.

Cleaning nets at Lowestoft after a prolific voyage, probably taken in the early 1920's.

CHAPTER FOUR

The Home Voyage

> *'Now up git the herrin',*
> *The king o' the sea.*
> *Says he to the skipper,*
> *"Look under your lee."*
> Chorus: *For it's windy old weather, stormy old weather.*
> *When the wind blow we'll all go together.'*
> (Traditional — 'The Haisbro Light Song')

Until the European industrial fishery of the 1950s and 60s began to pillage the nursery areas and damage the stocks irreparably, every autumn saw the concentration of vast shoals of herring off the East Anglian coast as they moved southwards towards their traditional winter spawning grounds in the Dover Straits and off the coast of Brittany. It was a migration that had been going on for well over 1,000 years and it seemed as inevitable as the roll of the seasons. The hauls made were prodigious, and not just by virtue of the numbers contained in the shoals. The actual configuration of coastlines in the southern North Sea may also have had a good deal to do with it, since their convergence towards the Dover Straits concentrated the fish geographically. Thus were the two ingredients for commercial success provided: huge quantities of herring and comparatively short distances to and from the fishing grounds.

Another factor in the importance of the East Anglian autumn voyage was the condition of the herrings themselves, which were in a prime state and suitable for every kind of processing. This excellence resulted from the species' feeding habits and resulting fat content. The shoals actually began feeding in March on tiny marine organisms called plankton (especially the copepod, Calanus), and as these flourished in ever warmer temperatures so too did the herrings' appetites increase. Feeding reached a peak in May and June, with a fat content sometimes as high as 25%, and then began to decline from July through to October. By the end of the latter month feeding had almost ceased and the winter fat content of the herring could be down to 5% or thereabouts.

The summer peak had its disadvantages in some ways because the fish were very, very oily and required extremely careful handling. Men who went on the summer voyages to the Shetlands and North Shields testify to this. Frequent references are made to the herrings bursting open simply because of the oil they contained. And in this condition, of course, they wouldn't keep for very long. There were no such problems with the East Anglian autumn herring. The fish had stopped feeding and the fat content was down to about 10%. In this condition it was perfect for most of the traditional ways of treatment. It was good for klondyking (exporting fresh in salt and ice to Germany); it bloatered and kippered well; it was ideal for the Scotch cure; it vatted favourably; and it made the most superb red herrings. As a result, the market for it was wide and varied — another factor at the time for the herring's importance in the respective economies of both Great Yarmouth and Lowestoft.

It wasn't just local men, however, who participated in the home fishing, as it was called. Every year saw a veritable invasion from the north in the hundreds, nay thousands, of

Steam Drifter's foredeck with — "Molgogger", Rope roller and Steam capstan.

THE HOME VOYAGE

Scottish fisher-folk who descended upon the two towns. This migration had begun to get under way in the last decade of the 19th century and it became a great annual event with the rise to pre-eminence of the steam drifter, the east coast ports of Banff, Buckie, Fraserburgh and Peterhead being particularly prominent. Whole families came down, the menfolk on the boats (which were often family-owned affairs) and the women on the railway, their portion when they arrived to work in the curing yards, gutting and packing the fish. The Yarmouth and Lowestoft landladies, of course, were glad of the visitors for about 10 weeks' extra income after the holiday season had finished, and many local families also took in these migrant workers, banishing 'spare' children temporarily to the attic or to grandparents — or even to the garden shed! It will be a long time yet before the Scottish invasion is forgotten in the towns.

Before the home fishing got under way at the beginning of October, there was always, traditionally, a short voyage made to fill in between the summer and autumn seasons. The boats returned from the Shetlands either in the middle of August or towards the end of the month (the Lowestoft men liked to be home for the town regatta, which was usually held during the last week) and then spent a week or more making up. After this period of overhaul and preparation they were ready to go down on the North Sea, as it was called. They spent about a week fishing out of North Shields and then worked south, having a night or two off Whitby and Scarborough. The voyage ended at Grimsby, with the boats getting catches on the Dowsing Ground off the coast of Lincolnshire, by which time the end of September was approaching and everything set for the great bonanza. The weather was of course very important to success and nobody knew more about it than Ernest 'Jumbo' Fiske (1905—1977), who was probably the greatest herring skipper of the 20th century:

'The best weather for fishin' on the hoom fishin' was after a good sou' west breeze, and then fall away. You know, drop away. Say a good ol' force 6 or 7 and then drop away. Drop away to about 2 or 3, 3 or 4. They used to stick their snouts in then and swim up then, they did. But on the real hoom fishin', on the full moon, that could be calm or anything, yit that allus seemed you got herrin'. On the October full, and November — anywhere about that time. Yis, you could git 'em in fine weather, except when that wun't very dark and there was a flat calm. You wun't git much then. No, you wanted a nice breeze, a nice steady breeze, a nice 4 or 5. Or, like I say, if you had a good blow through the day and then fine down, you were supposed t' git more herrin' then, you were. If you got gales o' wind, sometimes that was all right, but you did git a good lot o' spoilts. When the weather was blowin' right gales, you never got much at all. The ol' easterly wind weren't too good either, but I've seen some good shots when there's been a good ol' north-easterly wind all day and then fine down at night. I've seen some good shots then. Yis, I have. Some good ol' shots. Course, that generally used t' blow moostly durin' the day, then ease down at night.

'When we used t' go further up orf the San Ditty (The Sandettie Ground near the Dover Straits. In fishermen's parlance 'up' always means south, while 'down' indicates north) and them places, you wanted decent weather up there as well. Yeah, just moderate breezes like, you know. You didn't want no gales. Ol' northerly gales were the worst, ol' northerly winds. That was poor fishin' then. Yeah. You never got a lot out o' north-west gales neither — only hard work! Plenty o' spoilts and all that like, you know. When you were down at the

THE DRIFTERMEN

'Shooting' – the nets.

Mast & Gear and Wheelhouse omitted

Hauling

THE HOME VOYAGE

Shetlands, if you had a good stiff breeze, say force 5 or 6, you allus seemed t' git a good fishin'. That you did. But if you were there when that was flat-a-calm, you never got hardly anything. Sometimes the wind would be westerly or sou' westerly, sometimes nor'ad. You can't beat a sou' westerly wind anywhere for herrin' catchin', I don't think. Anything westerly or sou' westerly allus seemed the best direction t' have.'

What Jumbo Fiske had to say about the weather here is both interesting and accurate regarding herring behaviour. The fish always swim into the wind in the mid-water current, and there is a compensating flow in this layer which goes in the opposite direction to the wind at a faster rate of drift. Thus are the herring able to make way. Once this principle is understood, the accuracy of the observations in the previous two paragraphs will be clearly seen. A north wind would delay the shoals in their southward movement, whereas a southerly one would draw them down faster on the autumn migration. Again, the east wind was bad because it drew the herring away from the coast and dispersed the shoals, whereas a westerly or south-westerly breeze brought them in close. The whole business is a good example of what the fisherman knew by observation being confirmed by the scientists. A similar example was the signs that revealed whether there were herring to be caught or not. Frank Fisk (born 1899) gives an account of what the crews used to look for:

'Some ol' skippers, they'd look at the water. They could tell by the water, the colour of it, some skippers could. I believe one o' the Catchpoles was a master-man for that. Yeah, that was lookin' at the water; they could tell whether there was herrin' there or no. A skipper would always git a pail o' water and look at it. You know, he'd look at the water and if that was nice and green he'd generally shoot. Sometimes you could tell by the oil on the water. You know, the oil from the herrin'. Gulls would tell yuh as well. And blowers, porpoises, if you see any o' them about. And didn't they stink — pooh! — when they used t' come up, specially when you were haulin'. If you were haulin' where there was quite a lot o' herrin', say you'd got a shimmer o' 50 or 60 cran, time you were a-haulin' these big ol' blowers would come up and, oh, they'd stink yuh out! You know, you could really smell 'em. And then they'd go down agin and they'd keep with yuh all the time till you'd done haulin'. They used t' call 'em blowers, and if you see one o' them you could say, "Well, we'll stop and shoot here tonight." Yeah, that's what you used t' do. You could tell if there was fish there, if there was herrin' there. If you see one o' them, you could bet your life you were goin' t' stop and shoot. Oh yeah. Blowers. That's what we used t' call 'em. They're like porpoises, only they're bigger. Gret big ol' fish. You'd see 'em come up and go down. Up they'd come, then down they'd go into the water.'

Ned Mullender remembers some of the other signs of herring shoals:

'When you were herrin' catchin' you looked for fowl and the colour o' the water. Some o' these ol' skippers even reckoned they could taste 'em in the air, but I aren't goin' as far as that. I don't think they could. But the water'd often be milky and oily, and that meant herrin'. People used t' have the judgement of water. I had this chap along o' me once, Buzz Barnard, and he wuz a damn good bloke on water. He wuz older'n me and he wuz a damn good judge on a bit o' water. He used t' say t' me, "I don't know, I think we'll git some herrin' tonight." I'd say, "Yeah, so do I. I think we'll git a cran or two." "We'll git more'n a cran or two!" he'd say. He wuz a damn good judge too; he could tell yuh. Another thing you used t' do wuz look down at the screw, the propellor, t' see the thickness o' the water from that. Nine times out o' ten that was painted white, the propellor. They used t' put white paint on it so you could see how thick the water wuz. Sheer water weren't no good. That had t' be thick.' (Rudders were sometimes painted white as well. The idea was to provide a backdrop against which to judge the colour of the water. Good signs were a reddish brown in summer, denoting plankton, and a milky tinge in winter when feeding was over.)

'Herrin' swim up round about sunset as a rule, but sometimes durin' the day as well. When I wuz in the *Impregnable (LT 1118)* round about 1927, we went out o' here and shot our nets one day. Well, the cast-orf, a chap come from Tunstall by the name o' Billy Nicholls, he wuz in the wheelhouse along o' me. I set there readin' the newspaper what I'd bought afore we went out, and I spose that'd be about three o'clock in the afternoon. All of a sudden, I say, "Well, I don't know, William."

He say, "What do yuh mean?"

I say, "I think them buffs look a bit logie (low in the water), don't they?"

He say, "Yis," he say, "I think you're right."

I say, "I think we'll have a look at them." We'd put the nets over about midday, I spose. That weren't far from here, only about 18 mile or so. "Yis," I say, "That look like herrin' there, Billy. Call 'em out and we'll have a look." So the crew turned out and when we looked, blimey, we'd got about a cran in the first net! "Well," I say, "that's all right." Course, the crew had come up with no gear on. You know, just t' have a look on, as we used t' call it. So they weren't prepared t' work. So I say, "All right, slack away. Put that net back. Pay it over and we'll have a pot o' tea and somethin' t' eat and then we'll go t' work." We went t' work all right! We got over 200 cran time we'd done haulin'. So there you had 'em swimmin' up in the daylight, and that sometimes happened like that.'

How the skipper handled the ship while the nets were being shot is described now by Jumbo Fiske:

'You always shoot afore the wind. You know, with the wind astern or on the quarter. You lower your mizzen, and if there's too much wind you lower your mizzen right down. Sometimes there ent s' much, so you tie it up then, just tie a couple o' ropes round it and slack it out t' what you want. But you've gotta keep right afore the wind. (This was so the combination of wind and tide would carry the nets away from the boat as she moved ahead, thereby preventing the propellor from being fouled. The nets were usually shot from the starboard side.)

'Well, then after you'd shot, you'd pull your wheel over and let her come broadside and then you'd let about a rope run out (120 fathoms length), a rope and a half, and then you'd hold it and put the tissot on. You'd anchor the tissot on t' your rope. You'd have your rope, your warp, slack and you'd put it on your capstan. Course, there would still be some rope left down in the rope-room. On a new rope you'd put some old net round t' stop the tissot slippin' and there was a lot o' oil (creosote) work out of a new rope as well. Soon as ever you'd done and hulled your last net over, you'd pull the wheel over t' starboard and come broadside t' the wind and let her lay till you'd got as much water as you wanted. Then you'd catch a couple o' turns round the capstan with the rope and check the boat up so she'd come round inta the wind. Then you'd hold her. Next you'd put the tissot onto the warp with a rollin' hitch, throw the rope orf the capstan and let it go slack, and then everything would drop onto that tissot. Then you'd drive (drift) up and down. You were with your nets; they were layin' up ahid of yuh, you see. Soon as ever you'd done all that, you'd pull your mizzen up and hang inta the wind. You'd stop your engines then 'cause they'd bin just tickin' over as you shot. But if that was a bad night, you hatta roll over slow ahid t' take the weight orf in case you parted. That was when there was a gale.'

That's the skipper's view of things. Herbert Doy (born 1900) now explains the intricacies of handling the ropes and nets for the man on deck:

'The hawseman and the whaleman used t' shoot the net-rope, and the net-stower and the net-ropeman and the stoker used t' shoot the lint. The lint is for'ad and the net-rope is aft. The mate put the seizins on the warp, and that was a dangerous job too at times — especially if you happened t' git foul ropes. The poor ol' cook was down the rope-room watchin' the ropes run out and sometimes you'd git a foul un. You'd got a nice job on then. That'd git t' the mole-jenny, jam inta that, and then the mate was liable t' git pulled over the side 'cause he was standin' alongside. He was there reachin' out forward and hitchin' the seizin' onto the rope after that'd gone through the mole-jenny. The buffs were already on the net-rope, tied on b' the strops. One o' the men what weren't shootin' would go down inta the hold and chuck the buffs up. The strops had t' be passed over the hids o' the men who were shootin' the net rope or else they'd git pulled over the side. That was just the same with the seizins — they had t' be passed over the hid o' the man shootin' the back or else he woulda gone over as well. The cast-orf used t' pass the seizins t' the mate, so he'd be fore-ship as well.

'Sometimes you'd use your engines when you were shootin', but if there was enough wind you'd just use the mizzen. The engine come in when that was really fine weather, no wind. You'd shoot with it runnin' over slow. Some ships used t' pick up the nets an' all. That *Impregnable*, when I was in her, she'd pick a bloomin' net up quicker'n anything — in the propellor. The net'd swale round. That was the bind o' the propellor what'd drag it in. She was a devil for that, the *Impregnable*, because she was more of a flattish bottom boat. You'd mainly shoot the starboard side, but I have shot orf the port side an' all. That all depended if you wanted t' turn your net or not when you hauled. Yuh see, when you hauled, you'd want t' be the other way about if you'd shot orf the port side. That was why that was moostly starboard.'

Ned Mullender provides further details:

'When you shot, you had a net-roller fixed above the side o' the hold so the nets would come out easy. Otherwise, you'd hafta keep 'em up orf the coamins as you drew 'em to yuh. There'd be one man down the hold, shiftin' the buffs out o' the wings and he'd hand 'em up t' the men on deck. You'd have 3 men for'ad in the kid — one shootin' the lint, one the back-rope and one passin' the seizins over the hid o' him what was shootin' the back. They would be the net-ropeman, the net-stower and the stoker. Now aft you'd have the whaleman and the hawseman shootin' the net-rope. Then there'd be the cast-orf a-handin' the seizins t' the mate t' put on the warp. See, the mate would be there at the molgogger as the ropes were runnin' out o' the rope room. The rope run out through the molgogger and the mate'd stand up front and put the seizins on on the seaward side. The cook used t' be down the rope room watchin' the ropes run out, and as the bends and splices come up he would sing out, "Bend a-comin'!" or "Splice a-comin'!" The skipper would be in the wheelhouse and the driver down below. If there was enough wind t' shoot with just the mizzen, the driver'd probably be sweepin' the tubes or cleanin' the fire, or somethin' like that.

'The cork-line was shot aft and the back-rope was shot for'ad — unless you were shootin' orf the port side. See, you always shot orf the starboard side, except when the swell was aginst the wind. Well, if you were shootin' with the engine goin' in them sort o' conditions, the nets'd go underneath and catch in the screw. See? So consequences was there was certain times when you shot on the port side, though that was rare. That meant you had t' pass the seizins underneath the net. Not over the top — underneath. See? And o' course the fellers what was a-shootin' the net-rope, they had t' step over 'em. Course, you were workin' the wrong way round, yuh see. The net-rope was always stowed aft, so that had t' be pulled for'ad and you had t' put the seizins underneath. That was a bit complicated, but you could do it all right when you had to.

'When you shot, you pulled the nets to yuh in about three draws. You'd go "1, 2, 3," and that'd all be in your hand. Then you'd both swing together, and so would the other 2 on the other end, and you'd throw 'em away from the ship's side. I never went in a Yarmouth ship t' see what they did, but they always paid theirs over. I have known 'em t' pay some away on the Low'stoft boats. You know, one or two nets like, but not the whole fleet. When you shot the first buff and the first net over, you used t' say, "In the name o' the Lord, pay away!" That usually come from the skipper. (Traditionally a reference to the New Testament story of the miraculous draught of fishes.) I should say that used t' take about 20 minutes t' shoot, or within the half-hour anyway. Sometimes that used t' be quicker. Course, you shot more'n a mile and a half o' nets and you could say that you were shootin' at an average speed of about a mile and a half an hour. That was what the boat would be doin' — no more'n that. If you were shootin' with your mizzen and there was a nice breeze, you had t' scandalise — what they called scandalise the mizzen. That meant you dropped the garf (gaff) down so there weren't so much sail, and that would slow the ship up. Sometimes, if there was a nice fresh wind and you were gittin' the nets out a bit quick, a cork would sometimes foul in a norsel. You wouldn't have a lot o' time t' clear that then. I've seen 'em come tight and knock the whaleman or the hawseman over, and he'd been under the net, draggin' along. You'd go full astern then. See, he'd be under the nets as they were draggin' out o' the hold. He wouldn't go overboard; he'd go underneath and the nets'd be a-drawin' over the top of him. That weren't no laugh when that happened — more swearin' than laughin'. O' course, the skipper had t' to full astern then, and then afterwards he had t' git his boat up right agin for shootin' the nets.'

Once the shooting was over came the drifting along on the tide, with the boat secured to the nets and following their line of travel. Then, after a period of time, which varied greatly according to what conditions prevailed, the skipper would decide whether or not to "go to work". If there were enough herring in the stem end net and the buffs looked to be fairly low in the water, he would give the order to "turn out". Then, with a shout of "Busky-O!" or "Work-O!", the crew would begin the arduous business of hauling. Jimmy Fisher (born 1912) describes what the procedure was:

'How long you took t' haul all depended on what herrin' you'd got. I mean, with empty nets, three or four hours was about it. But if you got a good dollop, then you could be 10 or 12 hours a-haulin' 'em. When you started haulin' you'd be down in the hold, and you'd shake the nets from there. You'd be right down on the perks and the herrin' would drop down the gap near where you stood. Then you gradually built yourself up on the nets until you were high enough t' start shakin' the herrin' inta the kid. As the ship rolled, so they'd go down through the holes, and sometimes in fine weather you'd help 'em down wi' roarin' shovels. When you're haulin' it's 1,2,3, shake. 1,2,3 shake. There were four blokes in the hold and two in the kid. There'd be one on the net-rope and one on the back. The cast-orf would be for'ad takin' the seizins orf the warp and lookin' after the capstan. He adjusted the speed o' that accordin' t' the speed o' the nets comin' in. The man on the net-rope, he'd pull the buffs in and another man, him what was stowin' the net-rope, would chuck 'em in the wings. The cook would be down in the rope-room a-coilin' the ropes. That space was only about 6 foot by 10, and you used t' start right down on the bottom and coil up. There were wings down the side which they stored spare cutch in, and you had a board handy so that when you got so high up you'd ship that across from side t' side. Then you'd git up on that and carry on coilin'. If you made a hash o' coilin', so that you had t' coil some o' the warp up on deck, well then you'd got the job o' pullin' the ropes up agin out o' the rope-room and re-coilin' them t' git 'em all down below.'

Ned Mullender talks about the importance of teamwork in the hauling of the nets:

'There was an art in scuddin' as well. I mean, if you got one or two awkward fellers a-haulin', they could rip your hands out. You know, actually rip your hands. See, you didn't wear no mittens. The man on the cork-line, he might have 'em, but not the fellers what was a-scuddin' 'em out. And you was all shakin' the same time on that. It was like, say, dancers doin' the same step at the same time. You'd say, "We'll have 1,2,3, and then a shake." See, you'd draw the herrin' to yuh over the rail-roller and after 3 draws you'd got enough t' shake inta the kid. See? The kid would be about 4½ t' 5 foot across and the bunker lids would be out and the herrin' used t' run away below. Sometimes, if you had a big lot o' herrin' in your nets, you'd only have two draws and a shake. Mind yuh, everybody knew exactly what the other man was goin' t' do; that's if you were a goodish crew. But if you got an odd man what was a-comin' up as you were a-comin' down, you can realise what the cotton would do t' your hands!

'If you had an exceptional catch, you'd sometimes haul on deck. You'd put your hatches on. You'd got a big bearer go across the middle o' the hold and the hatches would fix on that. After they were secure you'd haul on them. All the nets would be on them too. You had a scuddin' pole t' help yuh work. There'd be an iron rod come up from the deck near the front o' the hold and then on the front o' the wheelhouse you'd have a cup thing. Well, the scuddin' pole fixed inta this iron upright at the fore end o' the hatch and inta the cup on the front o' the wheelhouse. That was about as high as your middle and you used t' lean aginst that when you was haulin'. You scudded the herrin' out inta the kid and stacked the nets on your side o' the hatch. You'd level 'em orf so they wouldn't git too high. Now if you could go about, you'd haul and stack up the other side as well. But if you couldn't do that, then you'd manage the best way you could. O' course, when you hauled like that, on the hatches, you'd git your mast up out o' the way. You know, your foremast; you'd git that up out o' the way. You'd throw the buffs down the fore-room. The fore-room hatch cover would be orf and you'd hull 'em down there. There, or on the after-deck.'

Hauling a fleet of nets was both tiring and complicated, we have heard. Jumbo Fiske remembers how it seemed when he began his career at sea and how the job was made slightly less onerous by the crew changing round tasks:

'I packed the farm up and went down t' Low'stoft and got a berth. I started as cast-orf, not cook. You soon larnt. Blimey yis, you soon got hold of it. The crew would show yuh. If you were willin' t' larn, and I'll always say this about the fishermen, they'd help yuh. Yis, they were very good. A couple o' hauls and you knew what t' do. I'd be on the fore-deck lookin' after the capstan, watchin' the rope dint ride, and takin' the seizins orf. Sometimes you used t' git a spell o' pullin' in the back-rope and that sort o' thing. And o' course, if you were gittin' right a lot o' herrin', the ol' chief engineer used t' come and lend a hand, or even the skipper. In the drifters you stowed the buffs down below, in the wings, but the seizins used t' be took orf the warp and pricked through an eye (loop) on the back-rope.

'That was hard work scuddin' when you got a big lot. A cran, cran and a half a net, is nice handlin'. But when you'd got two or three cran to a net, you'd pull, pull, pull — especially with them bran' new nets. They'd be a bit oily and slippery t' hold t' start with. There'd be six blokes haulin' altogether: one haulin' the back and one haulin' the net-rope, and then there'd be four down in the hold. A couple o' them would be right in the middle lint and then there'd be one either side o' them. The ones what had the hardest graft were the two right in the middle o' the lint. But o' course you used t' spell round. You swapped over. You didn't stay in the same place all the while. Oh no. See, if a bloke started by haulin' the net-rope, he'd haul his spell on there and then he'd come down and stow the net-rope. And one o' them out o' the middle would go up on deck on the net rope, and him what'd bin stowin' the net-rope would go on the middle lint. Like that, you all got a turn. The one stowin' the net-rope would put the buffs in the wings and the one who was on the back-rope would coil the seizins and look after them. He'd prick 'em through an eye, like I said before. The cast-orf was on the capstan and the cook was coilin' the ropes all the while. He used t' coil 'em right round. There was a knack in that. Ol' Alec Seamons, you could roll a marble on his ropes. When they were new they took some doin'. They'd keep turnin' up at one end, yuh know, till the tar and that wore out of 'em. That tar was hard on your hands; you used t' have t' git grease and rub on.'

During the home fishing many of the catches landed came from Smith's Knoll, a ridge in the bed of the North Sea about 30 miles north-east of Lowestoft. It was not only the most important local ground; it was, without doubt, the world's premier herring ground. Herbert Doy talks about "working the Knoll," as it was usually called, and elaborates further on the matter of hauling:

'Driftin' is hard work, harder than what trawlin' is. Cor, set light, yes! Everything is all hove up on winches an' that a-trawlin, but you'd gotta pull herrin' nets in. You git about a couple o' hundred cran o' herrin' and they want some pullin' too. (A wet, empty net weighed about a hundredweight. A cran of herring was 28 stones. A moderate haul of only one cran per net therefore meant a total weight of 40 stones.)

'Then you're gotta shake 'em out. You could be on that job eight or nine hours with a decent shot. That you could! You'd shoot just afore the close. That was the best time, that was. I've seen 'em wait for the tides when they shot at the Knoll Buoy, up here orf Low'stoft. They used t' wait for the half-tide, and then they used t' take what they called a half-tide down from the Knoll Buoy. Ol' Jute Burgess, he got a bloomin' load o' fish there. That's where he used t' work all the bloomin' time. Yeah, he used t' git a lovely voyage from there — the Knoll. He used t' work that buoy, the Knoll Buoy. That was the best herrin' ground, Smith's Knoll. But you'd got t' watch it because you could sweep the buoy. When you were drivin' down, your nets could drive through that. I've bin parted by the Knoll Buoy and gone and picked the other end up. You'd wait till daylight. You didn't go after them there and then, not in the dark, because you couldn't see. You'd wait till daylight. You'd keep driftin' along just the same and you'd see them a-driftin' down with yuh. Then you used t' go and pick 'em up, knot the rope up so you could haul. Once you'd done, you'd lay the ends out and start splicin' 'em. You know, splicin' the ends o' the rope.

'When you hauled, there was four of yuh down below, down the hold. Then there'd be two in the kid — one aft pullin' the net-rope in and the other one for'ad pullin' the back-rope in. The men down below stood on the perk boards and they'd be a-shakin' the herrin' out. They'd build all up on the nets and then they'd change round. Yes, you'd haul a quarter, a quarter of a fleet, and then you'd shift round so that you got a quarter of a fleet each. You'd go up on the net-rope and the man on the net-rope would come down below, and you'd work your way round till you finished up on the back, haulin' that. Oh yes, you'd all spell one another. The middle lint is the hardest work when you're gittin' herrin'. There used t' be two men in the middle lint because that was where the weight was. And that's why you used t' spell over — t' give each other a rest, see. You could git sore hands on that, but o' course you got used to it. You had t' all pull together and all shake together. Up you'd go and down you'd come. If they were alive they'd come out easy; but if they'd bin caught in the fore part o' the night and they were dead herrin', then they were harder t' git out. When they were alive they'd rattle out. That they would. Yes, you could allus git a live herrin' out better'n you could a dead one.

'If you got 'em double-swum (This was when the herrings, for reasons still not completely understood, swam into the nets from both sides.) that was hard work. They were on both sides o' the net then. You couldn't turn the net or do anything. I mean, if you got a shot where they were on the wrong side (where they'd swum in from the port direction), then you'd pull the net-rope for'ad so you were turnin' the net and scuddin' them out that way. When they were double-swum you'd gotta make the best of it. You'd gotta git 'em and then lift 'em over aft. You know, shake out the normal way and them underneath would drop out; but them what was on top o' the net you'd hafta throw orf, throw 'em orf aft. There were gaps between the perks for the herrin' t' go down, and you'd also have your bunker lids orf and shake 'em out into the kid. Then, for them what was left on top o' the net, you'd drop down aft. You'd shake them down t' the net-rope end and then throw 'em orf. They'd go down the after-well, which was just in front o' the wheelhouse. But that was damn' hard work, I don't mind tellin' yuh! Shake and lift all the way.'

One topic of regular interest in the driftermen's life was the Prunier Trophy, which was awarded each year by the proprietors of the famous London restaurant to the boat which had caught the largest single shot of herring during the East Anglian home season. It can be seen now in the Lowestoft Maritime Museum — a grey stone statue about a foot high, of a hand grasping a herring. Horace Thrower recalls how it was won in 1947:

'When I was in the *Patria (LT 178)* along o' Georgie Meen we won the trophy. 257 cran. I'm almost sure that's what we had. We come inta Low'stoft about 11 o'clock that night and there was a crew waitin' to unload the boat so we could go home. We were lucky t' git that shot that time, though. See, that was on the home fishin' and we'd started down on what we called the North Float — you know, just orf Cromer, that way. Well, we'd landed one mornin' and when we went back we couldn't git down onta the ground where we'd come from. So we tried short, just below the Cromer Lightship. We shot there. We never see another boat in the sea. That was dark when we started t' shoot and the skipper say, "We'll let 'em go till 12 o'clock." He say, "We'll look at 'em at 12 o'clock." So we did, and there weren't a herrin'. There weren't a sign of a herrin'. He say, "All right, we'll let 'em go till 4 o'clock." When we went for 'em at four, there they hung! I think there was one in every blinkin' mash when we first started. Cor, there was some blinkin' weight. You couldn't even see the buffs first quarter-fleet. They'd sunk an' gone down, so we hatta pull 'em up. As we were haulin' the wind shifted, so we hauled half one side and half the other. We got back inta Low'stoft at 11 o'clock that night and we'd gone t' work at four that mornin'. That'll tell yuh how long we were haulin' 'em in.

'Some o' the boats I was in had what they called a thief net t' catch herrin' what'd dropped out o' the nets. That used t' go out from the wheelhouse, or rather the pole what held it did. The net was like a bag, see, and you'd use it when you'd got a good lot o' fish. If they were droppin' out as you hauled, they'd fall inta the thief net. That'd be underneath the nets as you were haulin', hangin' up for'ad, and that'd catch what fell out. You rigged it so that it didn't foul the nets and the weight o' the herrin' pullin' on it made a bag of it, more or less. We didn't use a didall very much, but we allus used t' carry one. Sometimes we'd use it. If there was any quantity o' herrin' fallin' out and you didn't have no thief net, praps the skipper or the chief would stand aft with the didall t' catch 'em as they dropped. See, you'd git what you could, specially when they were makin' good money. I mean, they didn't make good money very often when I first went t' sea, but after the Second War they did. See, as the years went by, you got a better price for your herrin' each year. But the first years! I mean, sometimes that was a hard job t' sell 'em. Mind yuh, I was never in a boat what had t' dump good herrin', thank goodness!'

Of course there were many times when it wasn't all plain sailing. Jack Rose (born 1926) recalls some of the things that could go wrong:

'The nets used t' drift with the tide, and if there was anything in the way — well, naturally enough, somethin' was goin' t' give. I've bin parted by some obstruction when I was on watch and I've looked around me and thought, "Where the hell's all the fleet gone?" You'd look for'ad t' see if the nets were all right and there was nothin' there. You'd just pull up a bit o' the rope so many feet long. All gone. All parted. You'd gotta search round all night lookin' for 'em then. Things could be awkward when you were shootin' as well. The cook would be down the rope-room watchin' the rope snake out so it didn't catch. That was all one long length, with what they called bends and splices in. Where the rope was spliced, or where that was bent on, the cook would call out, "Splice a-comin'!" or "Bend a-comin'!" And that would give the mate and them a chance t' git ready. Well, one ol' cook, Stutterin' Freddie (the poor ol' boy used t' stutter), he used t' have a whistle. He'd give two blows for a splice and one blow for a bend.

'You could have some rum things happen when you got a good shot o' herrin'. The man on watch would have a look-out every so often. He'd go for'ad and he'd have a look at the buffs, and if he see them tippin' he'd know they were gittin' herrin'. Another time praps he'd go and call out one o' the other blokes, and they'd pull in a bit o' slack and have what was called a look-on. If you see herrin' comin' quick and heavy, then that was time t' git out and haul, else you'd lose the lot. The weight would just part your rope. That have happened as well where the nets have all sunk with the weight. You had a job on then. Dorgs (dogfish) were another thing; they'd mess the fishin' up. You'd pull the nets in and all you'd have was half-herrin'. The dorgs had bit 'em in half. You'd stand there shakin' your arms orf scuddin' out half-herrin' what weren't no good to yuh. What a way t' earn a livin', eh?'

Billy Thorpe (born 1908) remembers some mishaps too:

'Out on Smith's Knoll the skippers would weigh the tide up. They got so experienced in their knowledge that they could weigh the tide up t' miss the Smith's Knoll lightship and the watch buoy and also the Knoll Buoy. I've bin parted by the Knoll Buoy, though — misjudgement where the skipper shot. That's surprisin', yuh know. You've got your nets in the water and this buoy is comin' towards you at a speed of three or four knots, and you're tryin' t' steam round it t' git out o' the way. Then your swing would part and you'd got the job o' pickin' 'em up. Sometimes the wind would git round and the whole lot would come round behind yuh, (Commonly known as "Buffs up your arse".) so you'd gotta let go o' the stem end and steam round t' pick 'em up the other end. You used t' burn an ol' flare t' try an' pick out the marker buoys. Then you could say t' the skipper, "We've now passed the quarter bowl, the half bowl, or the three-quarter bowl" so he knew he was comin' towards the pole end. When you got about three or four nets from the end, you'd have a gaily chequered buff there so you knew you were comin' towards your end. Once you picked them up, o' course, you made the after end your fore end.

'If you'd gone out late at night, that used t' be like comin' across a town out there on The Knoll. Everywhere you looked there was lights. You can picture a skipper — he's gotta find a space t' shoot among that lot, and that's where the experience came in. You know, for him t' find a place t' shoot down. When you went t' work, they'd say, "Busky-O! Work-O!" On the hoom fishin', when the herrin' were inside, (close inshore) you could turn out here at 7 or 8 o'clock at night t' haul and you could be haulin' all that night. Then, another time, you'd be steamin' in and cleanin' your nets, and then you'd gotta land your herrins. That might be midnight afore you landed all you'd caught and then the skipper would sometimes say, "Run your nets down afore you go home." Sometimes you'd do it there and then; sometimes you'd come down early in the mornin' and do it. The nets had t' be run down or the heat would spoil 'em. They could even catch fire (due to internal combustion caused by the herring oil). The nets were stored in the hold. Halfway down the hold you had what you called perks — boards what you stood on — and you stowed your nets on them. You'd stow about 100 nets on the perks, say, and the herrin' used t' be down underneath. If you got too many herrin', you'd hafta run the nets up orf the perks because you wanted that space. I've run 'em down t' the wheelhouse sometimes, half a fleet or more up on the deck. I mean, you git 300 cran o' herrin', that's a lot o' fish.'

George Stock (born 1903) was a trawlerman for most of his working life but he went herring catching as a young man and remembers very clearly what he saw and did:

'I was never a real drifterman, but I've bin in the wheelhouse along o' the ol' skipper and he'd say, "George, can yuh see any gulls a-divin'?" Course, you know, if you see gulls a-divin', then there's fish there. Some o' the driftermen used t' go by the water as well. Sheer water was no good; they liked the milky water. Then there'd be the blowers. You allus used t' see ol' blowers and you'd say, "I bet there's herrin' about here." The October full moon was the time for herrin'. Oh yes. A nice moonlight night, not too calm, and you'd see the boys all go t' work, the herrin' a-shinin'. You used t' shoot just afore the close and on moonlight nights you could see the herrin' sometimes. You could look over the side and see 'em. That you could.

'Cor, there used t' be some boats out here. There'd be that many buffs floatin' around you often didn't know whose they were. Well, praps you knew afore dark, but afterwards you couldn't always be certain. Yes, many a time durin' the night you'd see ol' buffs come floatin' up your stern. I've come on deck — you know, when that was time t' take a watch — and I'd say, "Where are them there buffs what were on our stern?" One o' these ol' boys would say, "They've gone down. You're all right. They've gone down." They'd give 'em the pick, yuh know. They'd shove it inta the buffs. I've never done it myself, but some o' these ol' boys would soon stick it in. They'd puncture the buffs so they didn't come up and rove around your nets. Yes, they'd soon give 'em the ol' pick. That used t' have a sharp point and then a hook. That was a boathook, only we used t' call it a pick. Oh, you used t' git foul-ups all right, but some skippers didn't pay no regard. You know, some o' these don men. You'd say, "Skipper, there's some buffs here." You know, when you were shootin'. But they wouldn't take no notice. "Oh, turn a blind eye," they'd say. "Be like Nelson. Don't see 'em." What about that? They'd tell yuh t' look out and see if there's anything, and then you mustn't see it!

'There were times an' all where the local men used t' cut the ol' Dutchmen's gear. Course, you'd never shoot near a Dutchman if you could help it because you can't drive with him. He'd have the old rough nets and big wooden bowls instead o' buffs. His gear was a lot heavier than yours. Mind you, the Dutchmen would usually be well outside The Knoll, or even further across, but there were times when you got foul-ups. That's when you used t' cut. There weren't many Low'stoft men what didn't have a nice big Dutchman's rope for a swing-on. You know, that was t' swing between your first ropes and the capstan, where that went out from your bow to the first buffs. That's where your real warps were, your own warps. These Dutchmen's ropes were nearly twice as thick and some o' the ol' boys, if they thought one looked a good, new un, they'd cut it and have that as a swing. You'd gotta have a swing, and a tissot as well. The tissot used t' go on t' the swing at one end and hook inta the bow o' the boat at the other. That used t' take the strain time you were drivin' down.'

This chapter began with Jumbo Fiske's comments about the weather. It is fitting that he should have the final word, because he dominated the fishing for over 30 years. The local men will argue, as only East Anglians can, about many things — but you will find no one to dispute the proposition that Jumbo Fiske was the greatest herring skipper within living memory. His giant frame and matching exploits made him a legend within his own lifetime, and the stories told about him are legion. Once in the Gorleston Road Junior School, Lowestoft, about a quarter of a mile from where Jumbo lived, a history lesson was in progress and the teacher asked the class who was England's greatest sailor. Without hesitation came the answer "Jumbo Fiske". Perhaps Horatio Nelson would not have been too offended at that opinion of a fellow East Anglian mariner. So let Jumbo sum up for us:

'That used t' be a rum un on the hoom fishin'. We parted more'n once. You just used t' pick 'em up. Yeah, you could soon do that. Then you used t' git the wind "agin the law" sometimes. You hatta look out for that. Either you'd lower your mizzen down and heave up, or you'd let go and steam round and pick the other end up. Course, the nets'd be the wrong way round then when you hauled, so you'd hafta shift 'em round as they come in. Then sometimes you'd have the screw fouled b' somethin' — you know, nets or ropes. You used t' hafta git somethin' on it if you could. You used t' carry a knife on the end of a pole, and you'd cut it out wi' that if you could. Then you'd give her a full ahid and a full astern so she cleared. If you had too much in you'd hafta git back t' port somehow and have a diver go down. Oh yes, a diver'd go down.

'I should think Smith's Knoll herrin' were the best herrin' you could ever git. You could cure 'em, kipper 'em, everything. That was about 20 odd fathom out there. On the Dowsin' you'd git more of a maisie herrin' and that was the place where you'd gotta shorten your nets and your buffs up because there's a lot o' wrecks about. The Sole Pit was another good place, and the Well Hole. They were out orf Grimsby. Then you used t' git some good shots up at Cape Gris Nez and the San Ditty sometimes. Course, when you'd had a good haul, you had t' pull your nets up — especially if the herrins was oily — because the nets would heat. That was a job. You used t' run them up and down, pull 'em up half on each side o' the kid. You'd have half a fleet on one side and half on the other. Oh, you'd gotta look after the gear. Then you'd pull 'em down inta the hold agin, or stow 'em aft.

'When you were lookin' for herrin' you'd go by the water — and birds. You could see the ol' water was like ginger pop if there was herrin' about. Then you'd have a look t' see if there was any blowers, or the little ol' porpoises you used t' git. When you were out orf Low'stoft here, when you were orf The Knoll, you could shoot any time. Yes, that you could. There's bin good shots orf here in the daytime. I went out here one Sunday; I only went 16 mile and I never come out o' the wheelhouse. I took her out, and after we shot the stem end net they say, "The buffs are goin' down!" They were an' all — wi' herrin'. So we all had t' spar about and git t' work. Sometimes you'd have praps three or four boats gittin' plenty, yit the other boats weren't gittin' any. That's what it was like. You'd git fouled nets too, you'd be so close. Lots o' things like that used t' happen. You always got clear, though. Yes, you got clear. You'd kinda steam and pull one way or pull the other, you know. One boat'd shear that way and the other one would pull the other. You'd all help one another, like. Course, hoom fishin' time there'd be a lot o' Scotchmen up here an' all. They used t' leave some money in Low'stoft an' Yarmouth if they done well! They used t' buy all manner o' stuff t' take back. They were a gold mine t' the shops here. I've even seen 'em go hoom at the end o' the voyage with a pianner lashed down on top o' the hold. Yes, they used t' buy some stuff up here. That they did. Well, the shops relied on it; the fishin' was all that there was here at one time o' day.'

'Spur' Dogfish.

THE DRIFTERMEN

CHAPTER FIVE

Landing the Catch

'Here's a health unto his Holiness,
The Pope with his triple crown.
And here's to the price of nine dollars
For every barrel in town.'
(Traditional — 'A Lowestoft Toast')

With the catch stowed all speed back to port was a skippers main concern. The old law of supply and demand dominated the herring industry and the fluctuations in price, both natural and contrived, caused much frustration among the fishermen whose earnings were directly affected. "Catching the market right" was an art in which some skippers were more skilled than others.

Very small shots of herring were hardly ever run into port, but kept as overdays until a larger quantity (hopefully) could be got. These second day fish never commanded as high a price as the fresh ones, but were still a saleable commodity. It wasn't unknown for some herring to be kept until the third day in the event of poor fishing. They were then known as over-overs and were at the limit of safe keeping without the use of either salt or ice. From the 1930s onwards a number of drifters stored their catches on board in boxes (wooden ones at first, aluminium ones later), but the traditional way was to have them loose in the hold.

Until the railway came to Great Yarmouth and Lowestoft, the method of landing herring was for the luggers to ground themselves in shallow water and offload directly on to the beach. The main centre of operations in both places was along the denes, and the scene is depicted in a number of surviving 19th century paintings and prints. With the advent of the railway things began to change and the role of the Great Eastern Railway was crucial to development during the second half of the 19th century. Most of its installations along the quay at Yarmouth have now either gone or have been converted to other uses by the North Sea oil and gas companies, but a walk around the vast cobbled expanse east of the river gives some idea of the massive size of the operation. The mile or more of quay was ideal not only for the berthing of boats, but also for the efficient operation of the Scotch curing trade. Across the other side at Gorleston was a further extensive area of wharf for landing herrings.

Lowestoft had no proper harbour of its own until the 1840s, when Samuel Morton Peto bought out the bankrupt navigation company and proceeded to improve and develop its facilities with imagination and style. When he sold his interest in 1862 to the G.E.R. the pattern of improvement was continued, culminating in October 1883 with the opening of the Waveney Dock. This was the nerve centre of Lowestoft's herring trade for some 20 years, until the increasing number of Scottish drifters coming down each autumn necessitated the construction of a new basin next door. The Hamilton Dock, begun in 1903 and opened three years later, is a direct result of the rise of the steam drifter, which enabled more and more boats from north of the border, (especially from the towns along the Moray Firth), to come down for the autumn voyage.

Herring being barrelled on Lowestoft quay. A large proportion of the catches went to Germany and other export markets.

LANDING THE CATCH

When talking of herring, the cran measure is always regarded now as the accepted unit of quantity. Yet it was not always so. From the Middle Ages right up until the early years of the 20th century herrings were always reckoned in long hundreds and lasts, with 100 long hundreds making up the last. Originally the long hundred was 120 fish (fixed by a statute of 1303), but somewhere along the line it grew to 132, the extra dozen being known as overtail. This was the perquisite of the fish merchant, who was easily able to demand his due from the fishermen. When the catch was landed the crew members actually counted them out in warps (one warp was four herring — two in each hand), placing the herring in baskets, which were then handed ashore. In the event of big quantities being caught casual labour was often employed to help the boats discharge their cargoes more quickly. Evidence of this method of unloading can still be seen at Lowestoft, where the fish wharf is built in two stages. The lower landing, as it is known, was for people to stand on and take the baskets of herring handed to them from the boats before passing them on up to the floor of the market.

The large increase in the number of boats using both Yarmouth and Lowestoft in the early years of this century put greater pressure on the available berthing places at market and generally necessitated a speedier and more efficient way of unloading catches. It was now that the pulley on top of the steam capstan began to come into its own, providing a source of power with which to haul loaded baskets up from the hold. The foremast's boom now served as a derrick and, with a man ashore pulling and guiding the rope, the baskets could be run backwards and forwards easily. The particular kind of basket used was the quarter-cran measure, which was legalised in 1908, but which had been quite widely employed before that date. The word cran itself derives form the Gaelic craun, a barrel of 36 gallons capacity, but the Cran Measure Act established a volume of 37½ Imperial gallons. The actual weight of a cran of herring came out at three and a half hundredweight, so the quarter-cran baskets held seven stones of fish. Before the boxed herring came in their use was almost universal and led to the term 'cranning out' to describe the discharging of a drifter's catch. The market itself was a hectic place when the fishing was in full swing. This is how Jumbo Fiske remembers it:

'On the hoom fishin' you had t' land and then git out agin as quick as you could. There used t' be a lot o' pushin' and shovin' t' git up t' the market. Good God, yes! There used t' be so many boats, yuh see. Yes, you had t' git a rope ashore and git up as far as yuh could, wait till somebody had done and then git in. You used t' keep screwin' up on the engine. When you landed the hawseman would take the sample up t' the sale-ring. All the companies had their own salesman. Hobson's had a salesman, Low'stoft Fish Selling Company had a salesman. All like that. They all had their own bloke. When the Scotch boats come down they used t' have Scotchmen sell their fish. Then there'd be the buyers. They used t' like yuh t' keep your baskets well full. "Keep 'em full!" they'd be singin' out as you landed. They didn't last all that long, yuh know, the baskets. They used t' stretch. So you made a lot more out of a new basket than you did an oldun. Some of 'em used t' git right worn and they'd hold a good bit then. The buyers made their money out o' the boxes when they come in. I mean, when them boxes were well full they'd git funny weight.

'Low'stoft harbour was a bad harbour comin' in. I always liked Yarmouth better than I did Low'stoft; you could see what you were doin' there. You'd got such a strong ebb at

Low'stoft, see. You were all right comin' in on the flood, but when you got a strong ebb — oh dear! And when you got an ol' southerly wind you got so much swell. I never clouted the dump heads though, thank the Lord! Several of 'em did, yuh know. Yes, that they did. That was a bad harbour in the dark as well. I mean, that was all right on a westerly wind and that was all right on the flood; but when you got after the half-tide with an ebb and a southerly wind, then that was a bad harbour. Very bad. When that was dark you used t' come up on the pier lights and that was all right once you got inside. Gittin' out was all right. Yes, you could git out all right. That was easier. You could put her on how you liked goin' out. Mind you, when you'd got big boats they'd want a bit o' water under 'em. I mean, if they were draggin' the bottom that didn't matter what you did. You could pull the wheel and they wouldn't answer. Sometimes the weather was so bad you couldn't git out. That was when there was south and east gales. You'd lay in then. Best thing and all, wun't it? When you got a force ten you couldn't do any good in that. No, that you couldn't.'

Once in dock it was often very congested with all the boats trying to land their catch at once as Horace Thrower remembers:

'Down on that Low'stoft market used t' be a rum un at one time. Plenty o' times you'd be half a day before you could git anywhere near the market. In the end you used t' put a rope on the quay where you thought you could git up and you'd heave up on that. You'd wind up on the capstan and try and git up that way. You'd hear the ol' wood creakin' then. Yes, you'd be tryin' t' git up there with the engines goin' full ahid all the time. You'd git jammed in, yuh see. As I say, you'd hear the ol' woodwork a-goin'. That'd be creakin' and goin' ahid — especially on the ol' wooden boats. Well, plenty o' times we've knocked the rails in. You know, when we were heavin' up between the two boats. We'd keep a-goin' up and up and up. Well, somethin' had gotta give, hadn't it?

'When you were tryin' t' land you had t' work your engine. You had t' go full ahid sometimes t' squeeze up t' the market when there was a lot o' boats in. Them ol' wooden boats used t' creak and groan. The stanchions cracked on them more'n once. You didn't worry about that. They used t' git repaired when you went inta dry dock at the end o' the voyage. Yes, you'd go full ahid t' git up — besides havin' a rope on the quay and heavin' yourself up on the capstan. The ol' berthin' master used t' come along sometimes and tell yuh t' ease your rope up. If you didn't do as he said, he'd chop through your rope and you knew what you'd gotta do then — git out! There used t' be some games there, gittin' up t' the market. Oh dear, oh dear! Twist and turn, first one way and then another. Push someone out so you could git up a little further. That didn't do the engines no good neither because the boat weren't goin' through the water. I mean, you were practically still. And you were usin' the power and your steam would be gradually droppin' back all the time. I mean, you couldn't go makin' big fires up when you were in the harbour — not for just tryin' t' git up t' the market. Cor, there used t' be some language down there! There used t' be some swearin' and cursin' and goin' ahid. Specially from the skipper. His head would be out o' the wheelhouse and he'd be a-bawlin' at somebody. "Slack this up! Slack that up!" They wouldn't do it today. The blinkin' deckie would soon tell them what t' do.

'When you were herrin' catchin' they'd be about two or three deep on the market when you come in. You know, the boats. When you were landin' stem on, that was very awkward. See, you'd got other boats' masts in the way when you were swingin' the herrin' ashore. You were liable t' catch in them masts, you see. You'd gotta heave the baskets right up. Sometimes you'd catch the forestays as well. Oh, give me a broadside berth any time; that was a lot easier. Blimey, yes. Stem on was hard work for everybody, and a lot slower. You couldn't git on as quick as you could broadside. I mean, you could land a couple o' hundred cran o' herrin' broadside in the time that took you to land a hundred stem on. Oh yes, that all depended on what sort o' berth you got when you were landin'. If you got broadside, you could land in four or five hours. Specially if you got a good buyer who'd have everything ready for yuh.

'I know when I was in the *Present Friends (LT 89)* we landed twice in one day with herrin'. I wasn't in her long, just time I was waitin' for another boat. Fred Darkins was the skipper's name and we went out o' Low'stoft and shot. We weren't very far out and, blow me, if we weren't goin' inta Yarmouth just afore dinner! Well, we went aground on the Scroby (Sands). That'd come over thick, see, and we couldn't exactly see where we were. We weren't on there about half an hour afore the tide turned and we backed orf. We went inta Yarmouth and landed, went out agin about fourteen mile from Low'stoft, shot agin, and we were inta Low'stoft that same night at ten o' clock with another hundred cran. O' course, Yarmouth was a lot easier t' land at than Low'stoft. There weren't the crowdin' up there at Yarmouth. I mean, you'd got a good stretch o' quay there, where all them oil places are now. There must be a couple o' mile o' quay there, altogether. There must be — up t' the Haven Bridge. Oh yes, they'd land all the way along there. Mind yuh, you didn't use t' go in there very often, just chance time. Oh no, Low'stoft people never liked t' go there if they could help it.'

Arthur Evans (born 1901), was a fish merchant whose long career makes him ideally qualified to talk about what went on when the fish came to market:

'The main trouble down there used to be when the boats were trying to get up. You got 200 boats in the harbour (you've seen pictures of them, I expect, but you can't really gain any idea from that) and they all had their ropes on the various posts. Nobody wanted to sit there, particularly on a Saturday aternoon, waiting for hours on end without being able to get their fish up. They'd probably sold it, you see, but everybody was waiting. Then they'd start heaving one against the other and shouting, "Let go that rope! Let go!" And of course they wouldn't let go. Then one of the harbour masters used to come along with a chopper and chop the damn ropes. Some of these Scotchmen would have killed him for two pins. But of course he had to do it. They couldn't all get through. You can't get a quart in a pint pot, and that's what they were trying to do. Those that were unlucky would probably be waiting until midnight to unload.

'They had a sale-ring for herring down on the market. It's been pulled down now, what, about 20 years, maybe even less. You used to go in there and they would ring a bell for the sales to start. That bell would be ringing all day long for one boat, two boats, three boats. The only time they broke was for the lunch hour. I think they used to break from about a quarter to one till a quarter to two, something like that. And if boats came in then, they had to wait. But that was the only time. And if people wanted to catch trains and they were still short of fish, the price would jump like glory. There used to be three trays inside the sale-ring and the boats' samples used to be shot in them and the catches would be sold one at a time. There'd be a ticket on each one to say what boat and what quantity, and then after they'd been sold they'd be shot back into the basket they'd been brought in. The person who'd bought them would keep that sample and when the boat's crew started taking the herring out, if they weren't up to the sample, the buyer would refuse to take them and there'd have to be a re-sale. They'd have to put another sample out then, a fair sample, and the catch would have to be sold again. You see, the fishermen used to pick just the good ones out at times, and sometimes they used to get away with it. The samples themselves were always left to the end of the morning or afternoon and then they were sold. I believe the money went to some charity — the Mission to Seamen or something.

'Herrings were usually sold off one boat at a time, but when things were really busy they didn't always come up to the sale-ring. They'd be sold off to various buyers without being auctioned — especially if the trade was a bit bad. The salesmen would go to the different firms and say, "Can you take this trip?" That'd be firms like the Co-op, Sayer and Holloway, all the big people. They'd take whole trips at a time. Usually the hawseman would take the sample up to the sale-ring. He'd just shoot it into the sample tray and wait for the salesman. All the salesmen would be up on a rostrum. Yes, they'd all be up there, and then the sales would go off, one after the other. I never saw any real trouble down there — nothing that couldn't be ironed out. Mind you, you'd see one or two scraps now and again. A lot of the business used to be done by telegram down on the market, where nowadays it's all telephone. There was an office to handle telegrams down on the end of the market, the Herring Market. They used to pump them across to the main post office. The telegrams used to go underground in a vacuum tube to the post office on London Road. After that we used to have to run them over ourselves. The tube business is going back 50 years or more.

'Sometimes you'd get too many herring down on the market. I've even seen herring dumped while I was working down there. Mind you, if was usually not very edible. Another thing I remember was one afternoon, a Saturday, when a feller came in with 40 crans. Well, there were hundreds of crans already there standing around, so he couldn't do anything with his catch. And he said, "Well, will you give me five shillings a cran for them?" And this Scotch curer said, "Yes, I'll give you five shillings a cran for them — if you take them to Yarmouth. And he steamed out of Lowestoft harbour with 40 crans of herring at five shillings a cran and took them in to Yarmouth! Ten pounds for 40 cran and he went that distance! He should have got 25 shillings a cran by rights, but there was no demand for them that day.'

A bad market as the reward for all his hard work was only one of the many frustrations that the fisherman had to face in snatching a living from the sea. Others are recounted by Ernie Armes (born 1902), a man who spent nearly 50 years on Lowestoft fish market, working for a variety of firms. The only time he wasn't down there all year was between the wars, when he worked for a German klondyke firm, travelling the whole British coast, following the herring round from port to port. His is a knowledge unrivalled of the market and the way it worked:

'Down on the market, when a drifter came in, there would be a scutcher of herrin' shoved into a basket, into a sample basket, and the hawseman used t' take that up t' the sale-ring. Now the sales used t' start at eight o'clock and praps on early fishings, when the boats were all in, there'd be a line o' baskets with a little ticket on 'em — LT 49, 50 cran; LT 66, 100 cran; that sort o' thing. There'd be just the number o' the boat, the amount and the salesman's name on the tickets. And o' course everyone would have a look round at the samples and then away they used t' go inside. They'd be took into the sale-ring. In the sale-ring were three sample trays and there was a platform with three desks on as well, and they generally used t' use the middle desk t' sell from. The herrins were shot into the sample tray by the fisherman who'd brought them and the buyers would all crowd round. Away would go the salesman and he'd start a-shoutin' out, "How much?" He was a company salesman; he might have been with Peacock, or Hobson, or Richards, or the Fish Selling Company, or with one o' the Scotch firms. The Scotch companies used t' have their own salesmen come down and they used t' sell their own fish, see. The whole business used t' go along in bids and they'd be knocked down t' you if you bought 'em. That's how it was done and you could buy what quantity you liked. If you only wanted five cran, you just had five cran. If you wanted t' buy the lot, then they'd knock 'em down to yuh. Occasionally you got people pushin' the price up, but they didn't last very long as a rule. I mean, a lot o' the buyers used t' work a ring together. O' course they did! — specially up in the Scotch ports.

Things used t' git very tight down on that market, I can tell yuh. There'd be boats all the way up t' the quay and there'd be other boats a-stirrin' away, trying t' git in. Then there'd some a-trying t' git out. There'd be the harbour master, or the berthin' master rather, shoutin' his hid orf, "Let go o' that rope there!" And I have seen a bloke with a chopper chop the rope away. You know, the mooring rope. Old Jonesy and that lot, old Billy Stone, they used t' carry a bloomin' axe and they'd say, "Come on, let go! Let go!" And they'd chop the rope. That they would! See, you'd git a boat wanting t' come up and he'd be going ahead as hard as he could go. Then he'd have a rope on to a bollard and the other end round his capstan and be heavin' up on that. And crack! You'd hear the timbers go. You'd hear 'em creak. Yes, that you would.

'There used t' be some rum old boys down on that market. Crikey, yes. There'd be what they called the scranners; they were the blokes what used t' work for theirselves in just a small way. They used t' go round and buy the little odds and bits up. Yes, they'd go round buyin' half a basket o' herrin' orf a boat what'd got half a basket o' herrin' left aboard. They'd give about a tanner for 'em, praps less. And in the course o' the day, with the little odd bits they'd bought and the little odd bits they'd nicked, they'd rake together about a couple o' cran and go and sell 'em to a merchant. Yes, they were the scranners. There used t' be old Yarmouth Jack and old Gooderham, old Jasper and Musky, and old Jack Lay. They were proper old-timers, they were. And they always had enough money t' git drunk on. You'd see 'em up the "Triangle" night time a-singing like the dickens.

'Then there were the kids what used t' git down there — the boys. Yes, there used t' be plenty o' boys down there gittin' in the way. Gittin' in the way and gittin' buckshee herrin'. Half these small fish merchants, they relied on the boys goin' and pickin' up herrin'. That they did! I could tell yuh one or two firms (I shan't mention no names, though) what used t' do that. Thrippence a bag they'd give 'em. See, when the herrins were swingin' ashore, there always used t' be some fell out down on t' the lower landing and the boys used t' be down there a-pickin' 'em up. They'd git a bagfull and away they used t' run to a certain man, and he'd give 'em thrippence. Thrippence was a lot o' money in them days. Yes, the best bloomin' seats in the Hippodrome were thrippence.

'When things were in full swing, I've left home on a Wednesday night and haven't seen my missus till Saturday night. My little ol' boys used t' bring the tea down. The second boy and the oldest boy, they used t' come down and bring me a bottle o' tea. When I did git home I'd been down there so long that the wife had t' wash me. I just fell asleep, so she washed me. See, in the old days there were no regulations; that was boat in, boat out. Time the herrins were there the boats would be landin' 'em. There'd be news come through like, "So and so will be here around about eight or nine o' clock tonight and he's got about 80 cran." And you'd have to hang on. You'd go into the office and kip, have a snooze for a little while. Then, suddenly, away would go the bloomin' bell and you'd have to turn out. Mind you, a lot o' late herrin' were bought privately. See, the salesman would say to a buyer, "Well, she'll be here about nine o' clock and she's got about 80 cran. Will you have 'em at last bell price?" See, they wouldn't go up for a sale; they'd be bought at the price reached at the last auction. That way they saved themselves time and trouble.

'After this last war they got a regulation t' close the sale-ring. They even broke orf for dinner, which was a thing unknown before. Yes they broke orf for dinner and they held the last sales at eight o' clock. Nothin' was sold after that time. Afore the war that used t' be open all night! Yes, all night. By gums, that was one go. You never stopped — unless there was a gale o' wind so there was no herrins. And very often we prayed for that. Yes, you ask my missus, or ask anyone. Oh, I've bin down there all night. O' course, you wanted it t' be all go in one way so you could git the money. Afore the war I got £3 15s Low'stoft fishin' time. I worked for a German klondyke firm called Figena. The Germans paid about five bob a week more'n the English merchants and they also paid overtime rates, which the English didn't. When I first went with 'em in 1924 I got 1/3d an hour, and then later that went up to 1/11d. We didn't do too bad, but you had t' work to earn it. Blimey, I remember one time when I got about twelve quid for the week's work. That was a week's work as well. I thought I wasn't goin' t' be poor no more! That was workin' all night and workin' Sunday. We worked one Sunday till four o' clock in the afternoon. They got special permission t' land herrins here because, as you know, the market was usually closed on Sundays. Just over twelve pound we took. My missus bought the kids some new clothes and that sort o' business. That was workin' seven days and nearly every night till past midnight.

'In the middle o' the market was what was called the subscription office. That was where a merchant who'd come down from, say, Scotland and had no office t' go to could find himself a place. He'd have just a little desk in that office on his own. There were about 30 desks in there altogether and some local men who didn't have an office also used t' use it. They paid so much a year to the railway comp'ny. That was a rum old game buyin' herrins, I can tell yuh. A real cut-throat business. Yes, the buyers were a sharp lot. One o' their favourite tricks was t' git a re-sale and git the herrins cheaper. That all stopped when the arbitrator come in just afore the last war, but up till then there were some rum tricks. See, the sample used t' come ashore in a little sample basket and that'd go up t' the sale-ring and be shot in a little tray and the salesman used t' shout out, "Now what can I ask for these?" And there'd be everybody lookin' at them and pullin' them about. They knew exactly what was there, whether there was any spents among 'em or whatever, and then eventually someone would buy 'em. The sample used t' have to go back with the hawseman who'd brought it up and that used t' stand near the side o' the ship till you'd done unloadin'. You'd be workin' away and then somebody would come along and say, "Bloody herrins have got t' drop. They're only makin' so much now." Praps yours had been bought at 28 or 29 bob a cran, but they'd now dropped three or four shillins. You'd git tipped orf, yuh see. Then along would come the buyer and he'd say, "Skim 'em round. Pick some bad uns out." So that's what you would do. Yes, you'd git hold of a few what were scaled or had dorgfish bites in 'em. Then you'd say t' the fishermen, "Hang you on with them herrins. They aren't like your sample." Course, they'd want t' know what was up then, so you'd show 'em the bad uns you'd picked out. "These aren't like your sample," you'd say. "You've got a good sample there." They'd have a look and then the buyer would refuse t' take any more, what was known as chuckin' up.

'Well, the remainder o' the catch would have t' go back t' the sale-ring then and there'd be a re-sale. O' course, they never fetched so much the second time, so you could buy 'em cheaper. Mind yuh, you had t' pay the original price for what you'd already took out o' the boat. The poor ol' fishermen, they were robbed right and left. I mean, that chuckin' business was bad. There was one bloke used t' be down there and whenever we saw him we used t' say, "Here come Steve Tucker, the champion chucker!" He bought for a German firm called Franz Witt and he was allus chuckin' 'em up. He walked about like a bloomin' corpse, poor ol' boy. He died o' consumption in the end. Yeah, he looked like bloomin' death warmed up walkin' along that market. The only time he come alive was when he was chuckin' herrins up. Mind yuh, he weren't alone. Some o' the Scotchmen were beggars for that. And the poor bloomin' fishermen were allus at the tail end.

'That was a hard life all round, right from the catchin' to the workin' of 'em. Workin' herrins on the market, you were stuck there all hours o' the day and night, hail, rain, blow or snow. That easterly wind used t' nearly blow yuh out o' the back o' the market! Cor, that didn't half used to drive through there. Another thing on the market, if the wind was a little bit easterly and the drifters were all coalin' up ready t' go and the water was low, all the smoke would blow under the market roof and you'd git as black as the bloomin' ace o' spades! Laughs! There used t' be some bloomin' laughs down there. Dear, oh dear. The first thing we used t' say to the fishermen when they started landin' was, "Now what about the gorger? Where's the gorger?"

'And the mate would say, "Gorger! No bloomin' gorger. Not yit. Let's git a few bloomin' herrin' out first."

"No," we used t' say. "You git the gorger out first." And, if you were lucky, a gorger o' tea used t' come ashore. And you could always tell a good cook from that. If the gorger was clean he was a good cook. But that often used t' come up full o' herrin' scales. And if you got tea from one o' the Scotch boats up in the Shetlands, that often used t' taste o' paraffin oil because a lot o' the old Scotch boats used t' run on paraffin.

'There used t' be one ol' cook out o' Low'stoft who used t' say, "I've made yuh a bit o' cake. I'll bring yuh a bit o' cake up." And one day I didn't want it. That was a nice bit o' cake, but I didn't want it. He got all ikey. "Shan't never make you no more," he say. I remember one boat, the *Marshal Pak (LT 200)*, and the herrin' were comin' ashore and I say, "What about the bloomin' gorger? You re a nice lot, you are. Nearly 90 cran o' herrin' out and we don't even git a drop o' tea."

'One o' the blokes who was haulin' the herrin' ashore say, "Tea! Tea! You'll be lucky t' git any bloody tea here. And even if we did bring it ashore, I bet you wouldn't drink it."

I say, "Why not?"

He say, 'We've got the dirtiest bugger a-goin' for a cook. do you know what?" he say. "He made a bloomin' beef puddin' the other day and he wrapped it round with the bloody 'Daily Mirror'. When we undone it, there was a photograph o' bloody Hitler on the puddin'!"

'Now that's the truth, that is. Cor, there used t' be some games down on that market. You'd be tryin' t' diddle each other all the while. "Keep them bloomin' baskets up!" you'd say t' the fishermen. "Keep 'em full."

"All right, mate," they'd say. "We'll knock you orf a few at the end."

"Never mind what you'll knock orf at the end o' the shot. We want 'em full now." So what they'd do then is put a full basket ashore and didall the herrin' out into the other baskets. There used t' be four baskets t' the cran, so you'd wait for the whole four t' come ashore before you let the fishermen sling 'em back down the hold o' the boat. They used t' chalk up the crans in what they called a gate on one o' the posts what held the market roof up. You know, that was 1, 2, 3, 4 and then across (1111). That'd signify five cran. One gate was five cran. And then, when they'd landed the whole lot, they used t' reckon up — 5, 10, 15, 20, 25, 30 and so on. That was how they used t' keep tally. If you didn't watch 'em, they used t' try and work an empty basket into the four. You'd only had three full ones really, but they'd go and chalk up a cran on the post. They'd try and tell yuh at the end o' the shot that you'd had 60 cran out, so you would count your klondyke boxes and say, "We bloomin' well haven't you know." (These were large wooden cases in which the herring were shipped across to Germany in salt and ice. They held 12¼ stones of fish).

"Yes, you have," they'd say. "Come and count." But o' course they'd been whippin' in that empty basket every so often. Oh, there were all sorts o' crafty tricks. I mean, we used t' try and rub out a gate from orf the post if we could. Hell if there weren't some language then.'

A CRAN — four baskets.

CHAPTER SIX

The Share System

'At his word the Ocean yieldeth
Bounteous harvest without fail;
His the hand our life that shieldeth
Through the fury of the gale.
Let us thank him,
Let us thank him,
For his mercies never fail.'
          (The Corton fishermen's hymn)

Drift net fishermen earned their money the hard way. Of that there is no argument. The living they snatched from the sea was both dangerous and uncertain for their earnings were as erratic as the North Sea weather. The method by which their rewards were calculated was a complex and mysterious affair and had its origins back in the dim recesses of the past. Traditionally the British fisherman all round our coasts was paid on a share basis. In other words, he received a proportion of the boat's earnings after all expenses had been met; and if, after settlement of these, there was nothing left, then he received nothing for his labour. At least that was the theory of it. By the 20th century things had begun to alter somewhat, particularly in the sphere of trawling where the crew (apart from the skipper and mate, who remained on a share) received a regular weekly wage. It was this security that prompted many of the Lowestoft men especially to go in for trawling, leaving drift net fishing to the men from neighbouring coastal villages and the hinterland, where many of them probably spent part of the year working on the land.

Though the steam drifter heralded a new era for the fishing industry it did not remove the share system of payment, which went back centuries and was simply modified to meet new requirements. For instance, with the arrival of steam power there was the specific need for engineers to drive the boats and look after the machinery. Such men were at a premium in the early days before the First World War and this is no doubt why they alone were paid a weekly wage while a fishing voyage was in progress (30 shillings) and also while the boat was laid up during the winter lull (20 shillings). During the latter peiod they would be expected to carry out all sorts of maintenance work, but at least there was the security of a regular income over a period of weeks when all the rest of the crew had was a few odd shillings made from the sale of gillded herrings when the nets were cleaned after landing. There were also a number of skippers who used to give the odd cran or so to the men to sell for themselves, thereby technically defrauding the company which employed them.

The share system of payment forced the crews to indulge in all sorts of minor fiddling, simply because it kept them waiting so long for their earnings — and kept their families waiting as well. There were three main voyages in the driftermen's year, the Westward journey round to Devon and Cornwall in the early months for herring or mackerel, the summer season down at the Shetlands, and the home fishing in the autumn and early winter. Each of these had its own individual settling, when the men were paid off with a share of what had been earned. There were other voyages as well, but they were not the general rule and so many men were faced with slack periods when their boat was laid up and when they were temporarily out of work. At such times some of them might be lucky enough to get in a few trips trawling, but this was not always the case because there were always more men than jobs. The best kind of vessel to be on was undoubtedly the drifter-trawler because this class of boat often managed to keep working all the year round.

One can imagine the hardship endured by families before the First World War, when all the basic necessities of life had to be obtained on the slate. The grocer, the butcher, the hardware merchant and the rest had to be owed money until the end of the voyage, when all debts were settled. Well, not all, because in the event of a bad voyage, with little or nothing to come in the way of payment, then credit had to be got for a further period of weeks in the hope that next time there would be a pay-off. It was a trying way of organising the family finances, and it was hard on the small shopkeeper as well. There were several of them in Lowestoft (particularly down on the Beach Village, where there was a great concentration of fishing families) and Yarmouth who had to wait a long, long while to get their money in and who often had to write off bad debts in times of hardship. After the First War things improved slightly because all crew members were able to draw a weekly allotment according to their status on the boat, which could be collected by their wives every Friday from the company office no matter where their husbands were fishing. In Lowestoft the concourse of women and children in the area of the Fish Market on a Friday was so dense that the whole affair became known as the market races. The money drawn each week was deducted from the men's earnings at the end of a voyage, so in the event of a bad paying-off they could end up in debt to the firm that employed them — a debt, one should say, which was never collected.

This was not so much generosity on the part of the vessel owners as an acknowledgement perhaps that they couldn't exploit their employees too shamelessly. Without a doubt the share system operated very much in favour of the owners. In Lowestoft they took 62½% of a drifter's nett earnings as opposed to the crew's 37½%, while in Yarmouth the proportion was 55½% to 44½%. Just why there was this difference between the two places is hard to say, but it may have something to do with the fact that in the era of the steam drifter Lowestoft had more big companies than Yarmouth — and big companies, as everyone knows, have more overheads and expenses than small concerns. And it was an expensive business, the purchase, fitting out and running of steam drifters. As long as things went well there were good profits to be made, but given a run of bad seasons and poor markets, such as occurred throughout the 1920s and 30s, then the fishing companies were in trouble — and this is clearly shown by the number of them which went out of business.

So complex was the share system that one finds comparatively few fishermen who actually understand fully the way it worked. Whether they were deliberately kept in the dark is debatable, but they certainly don't have a detailed knowledge of the way they were paid, just a general idea of the principle. A drifter laid at a certain number of shares per £100 of nett earnings, with the boat owner taking a certain number and with the rest divided among the crew. The nett earnings were the profits made on a voyage after all running expenses had been met — items such as coal, lubricating oil and engine stores, food, landing dues, shore labour, baskets, fish salesmen's commission, water charges, salt and ice. When all of these accounts had been allowed for, then the paying-off could take place.

Fishermen's cottages at Lighthouse Score, Lowestoft.

As we have seen the expenses sometimes exceeded the total earnings and that meant no pay-off for the men.

In Lowestoft a steam drifter laid at 24 shares, with the owner taking 15 and the crew 9, and this meant that the full share was worth £4-3-4d per £100 of nett earnings. At least that was the theory. In actual fact the Lowestoft share this century for driftermen was £4-2-6d, which meant that the owners were taking out the other 10d on top of what they were already getting. If an extra crew member was employed, as sometimes happened on the larger drifter-trawlers, then that brought the full share's value down again to somewhere around £3-19-0d. In Yarmouth the owner took 9 or 10 shares (it depended) to the crew's 7, which meant that the full share per £100 of nett earnings was worth either £6-3-4d or £5-17-8d. The crew's proportion was then divided into 10 smaller shares and their payment worked out accordingly, which meant they received something in the neighbourhood of £4-6-4d or £4-2-4d per £100 depending on the owner's proportion. Crew shares were then apportioned on the basis of rank, working from the skipper downwards, and operated on a system of eighths, or half-quarters as they were generally known.

In Lowestoft the arrangement was as follows: skipper, one and a quarter shares (10 eighths); mate, one share and a half-quarter (9 eighths); driver, one share (8 eighths); hawseman, one share (8 eighths); whaleman, three-quarter share and a half-quarter (7 eighths); netstower, three-quarter share and a half-quarter (7 eighths); stoker, three-quarter share (6 eighths); three-quarter man, three-quarter share (6 eighths); cast-off, half-share and a half-quarter (5 eighths); cook, half-share (4 eighths). When added up the total comes to 70 eighths, which left a quarter-share (2 eighths) to be used at the discretion of the skipper. The common practice was to give this to a man who was a competent cook, rather than take a chance on the rough meals served up by a young lad on the basic half-share. Yarmouth's system was similar, but there were variations: skipper, one and three-quarter shares (14 eighths); mate, one share a half-quarter (9 eighths); driver, one share and a half-quarter (9 eighths); hawseman, one share (8 eighths); whaleman, one share (8 eighths); netstower, three-quarter share and a half-quarter (7 eighths); stoker, three-quarter share and a half-quarter (7 eighths); two younkers (some boats only had one) a three-quarter share each (6 eighths); cook, half-share (4 eighths). The Scottish ports again had their own system.

As far as the owners were concerned, the sum total of their shares was a composite affair calculated on the capital and consumable costs of operating the drifter itself. The engines, the nets, the little boat and the capstan, etc. were all allowed for. Taking out a share for the capstan was a throwback to the 1880s, when steam capstans first began to appear and were seen as a new large item of capital expenditure. The fact that the owners continued to allow for this right down to the end of the fishing is always taken by ex-driftermen as being indicative of what a fiddle the share system really was. And there were others as well. A particularly ingenious one was the sticky-bag, which was occasionally used by unscrupulous owners to cheat young and illiterate crew members 70 years ago and more. The paying-off in those days was made in sovereigns and these would be put out (not counted) on to a table in front of the recipient so as to show him his due. They were then scooped into a small canvas money-bag, the inside of which was thickly smeared with tallow, before being poured out on to the table again. Any coins that stuck fast were kept

THE SHARE SYSTEM

by the owner!

One is bound to admit that this method of deception has a definite air of comedy about it. Generally, though the driftermen's lot was nothing to laugh about. The 1920s and 30s were especially hard times and what made it worse was that men who were out of work couldn't get dole money. Because they were share fishermen they were, in the eyes of officialdom, part-owners of the vessel on which they worked and therefore not entitled to dole. Health insurance was taken care of by the Lloyd George stamp, but there was no unemployment benefit. If a drifterman was out of work he had to apply for relief, prove near-destitution (which often meant first selling off family possessions) and then draw his handout mainly in the form of food vouchers. A man and his wife received 18/- a week, with 3/- for the first child and 1/6d for each subsequent one. But that wasn't the end of it because the relief had to be earned. In Lowestoft, during the 1920s, for every 7/6d worth of relief received a man had to put in a day's labour either on building the sea wall or doing test work at Oulton Workhouse, the latter usually consisting of gardening and other odd jobs around the place. Billy Thorpe remembers some of the ups and downs of those days:

'You went as cook first of all at a half-share, then you went cast-orf at half and half-quarter. In them days you never said five-eighths, always half and half-quarter. Next you went three-quarter share. You never used the word deckhand aboard a drifter, you see. Then the next berth was three-quarter and half-quarter on the after-deck. That meant you kept on the after-deck and was in charge of the mizzen. Well, then you had the whaleman, who was the same as the three-quarter and half-quarter, only he was on the fore-deck. And when you landed the herrin' he allus used t' be on the quay with the hawseman and the mate seein' t' things there. The hawseman was like a third hand, but you never mentioned third hand on drifters, only on trawlers. The hawseman came after the mate. The engineer allus used t' be called the driver because he used t' drive the boat, and the second was called stoker. They call some o' these chaps engineers what are on the trawlers now, but if the engine was t' stop at sea some of 'em wouldn't be able t' start it agin! A driver got a full share, so did the hawseman. The skipper got a share and a quarter, the mate a share and a half-quarter, and the stoker a three-quarter share.

'When you went t' sea you got a weekly allotment time you were fishin'. That wasn't exactly a wage and your wife used t' go after it every week because the comp'ny said they couldn't afford t' post it! I got married in 1932 (the worst year on record for unemployment) and I went mate o' the *Constant Hope (LT 32)*. I got 22/6d a week and they took 1/5d, I think, out of our money then. That left the wife with 21 and somethin'. The 1/5d was for health insurance. I borrowed £310 t' buy a house and everybody told me I was mad. I had t' pay 10/8d a week back t' the buildin' society and that left my wife with just over ten bob t' run the house. At the end o' the season you were hopin' t' git a pay-orf, but many times there was nothin' t' come. On paper even the comp'ny lorst, but they used t' git a share for the small boat, a share for the capstan and so on because a boat laid at a certain number of shares. When you paid orf in debt that looked as if the owners were losin' money. They'd say, "Well, we've bin givin' you a pound a week and now you're payin' orf in debt." Like that, people might think they were givin' you somethin'. But they weren't really; they were takin' it out in the other shares.

'On my best voyage I can allus remember havin' 60 pound notes — 60 bran' new pound notes! And I can remember now takin' them t' my brother in hospital t' show him. And he said, "Cor, let me hold them!" I think that was the most money I ever did take up, that 60 pound. That would be in 1928, the end o' the hoom fishin'. Then things began t' drop after that. You know, 1929, the thirties. They used t' call some o' the ol' boats white elephants. The ol' *Meg (LT 316)* an' the *Covent Garden (LT 1258)* an' them sort o' boats, you know, which never earnt a lot o' money. They weren't lucky boats. Some boats were. That didn't matter who went in 'em, they got the fish. When a drifterman was out o' work he had t' go on the relief. If there weren't no work they used t' send yuh t' what we called the workhouse and you'd have to put in so many hours up there. They didn't give you any money; you used t' git food chits. You used t' help in the garden up there an' give a hand wi' some o' the inmates too.'

Jack Sturman, born in Oulton Village, not far from the workhouse, also remembers the life of a drifterman in good times and bad:

'A boat laid at so many shares and when you went t' pay orf somebody from the comp'ny would read out all the expenses. Well, one o' the blokes would be sure t' pick on somethin' an' say, "What was that for? We didn't have that." O' course, there used t' be all sorts o' fiddles an' that. At one time we were paid orf in sovereigns. I've seen the time that a chap has come in the "Red House" at Oulton time we sat there an' say t' the landlord (the landlord had bin a fisherman), "Give all the boys a drink." He was well away, yuh see, an' he put his hand in his pocket an' pulled out a handful o' sovereigns an' he say, "There you are. An' if that ent enough t' pay for it, there's another lot there!" The landlord counted it all out an' there was 84 pound! Well, he took f' the round, but the bloke had gone by then so he picked all the rest up and counted it out in front of us what sat there. O' course at Chris'mas time, when you paid orf, there was generally plenty o' gipsies about an' there was one or two old boys sat there that night. Anyway, the landlord say t' us, "You see what there is?" Then he called his wife an' said, "There you are, mother, take that through inta the kitchen." Next mornin' the chap come down an' he was in a helluva flap. He'd lorst all his money! Well, o' course, the landlord let him keep on for a little while an' then after a bit he say, "You'd better go an' see mother down in the livin' room." See, if he'd took all that out with him when he left the pub the night before, like as not he'd have lorst it. The gipsies would have laid wait for him. Oh yes.

'Most o' the time there was a lot o' the boys out o' the country what went fishin'. They'd come from Thorpeness an' Aldeburgh way even, and o' course you got 'em from Winterton way, from Aldeby, and from Wenhaston and round about there. Several of 'em would just do the hoom fishin'. They wouldn't go any other time, but work on the land instead. O' course, I've known a good lot o' the local men, even the skippers, t' go stone-pickin' on Low'stoft Beach because they hent earned no money. There weren't no social security then. Some of 'em used t' go to Lothingland Workhouse down Union Lane at Oulton (course, I was born not far from there), and they'd go there t' work on the fields an' that like. They dint git any money, but they used t' git some groceries. You got a chit an' the only way you could git a smoke was if the grocer let yuh have a packet o' fags instead o' groceries. Some would work it in for yuh, but you only hed so much t' draw. Like I said, some o' the skippers used t' go stone-pickin' on the beach 'cause they hadn't earnt no money. Course, you'd sometimes hear people say, "Coo, he's done whoolly well!" That'd

THE SHARE SYSTEM

be the hoom fishin' time, when you paid orf. But praps that'd be the only money a bloke earnt all year. If that hadn't been for the grocers, the tradesmen, lettin' people have things on credit, a lot of 'em woulda starved.

'There were times when you paid orf with nothin'. Or praps you'd have a good hoom fishin' an' that'd be the only money you'd make all year. Sometimes you had just a few pounds t' take up and o' course that'd leave yuh in debt with the grocer an' that like. Some shops used t' let people have credit for the whole year. Now I was a driver an' that meant I got a full share, which was four pound and a half-crown out o' the hundred. Yes, I got four pound and a half-crown an' that was between the wars. Durin' the Second World War the drivers what were fishin' got two quid a week allotment instead o' the thirty bob. Some of 'em even got fifty bob. Yet after the war, when they all started orf agin, they were willing t' go t' sea for thirty bob agin. Course, in the real old days there weren't no union an' that like. When you paid orf, the owners would work out how many hundred pound profit there was an' then they'd know how much money t' pay yuh. The best pay-orf I ever had was about eighty pound.'

Although he spent his life on the market Ernie Armes was the son of a drifter skipper and remembers how the share system affected family life:

'That was hard when I was a boy. We never had no bloomin' money. See, your fathers didn't git nothin'. There was no such thing as a weekly payment. A man who went driftin' in them days, he only got what he earnt, an' if he didn't earn nothin' he never got nothin'. That wasn't till after the First War that they give the men a small sum o' money per week, but before that there was nothin'. Course, the engineer got a weekly wage, but that was knocked orf his share. Everybody was hard up at that time more or less. You had t' live orf what they called the book till the ol' man paid orf. If he'd done well there'd be two or three piles o' sovereigns on the table an' then my mother would git the bloomin' book out. There'd be Mrs. Robinson the greengrocer an' Mills the grocer an' all that. The ol' man would sit there an' he'd say, "Now then, Sarah, how much so-and-so? How much the rent?" And my mother would say, "Well, the book say so-and-so." Away he'd go an' count the money out. When they'd bin through all the bills my mother would praps say, "I had t' buy George a pair o' boots an' Ernie a coat." More money. "Bloody hell, gal!" he used t' say. "That's all bloody gone!" But he used t' keep enough so he allus got well drunk for about two or three days. They used t' go on the booze, all the fishin' boys; go on the booze for about two or three days.

'Yes, they'd all meet an' have a good booze up. Blimey I've seen 'em drunk at 8 o'clock in the mornin' when I was on the way t' school. When the hoom fishin' was over all the crews got stood orf, but Tom Thirtle, who my father worked for, he used t' keep his skippers on. They'd be down the beatin' chamber, ransackin' nets an' all that sort o' thing till the Scotch voyage started. They used t' go t' work at 6 o'clock in the mornin' — well, the pubs opened at half-past five! Away they'd go into the "East of England" instead o' goin' into work and they'd sit there havin' rum hots an' dickens know what. Then the ol' man would come hoom drunk for his breakfast. That was a common thing. "Oh dear, here he comes," my mother would say. "There's Tin Tack, Dirty Neck an' Cronjie on the booze again." Tin Tack was my father, Dirty Neck was Arthur Ladd an' Cronjie was Cronjie Capps. They were all Tom Thirtle's men. Arthur Ladd was a bloomin' gret bloke an' so was Cronjie Capps, an' my father was short. So there'd be Tin Tack in the middle! Yes, they'd soon git drunk. A rum hot in them days was about three ha'pence; a pint o' beer was tuppence.

'My father allus earnt a bit o' money, but the share system was a twist from start t' finish. A boat laid at about 24 shares and the crew's total came to about 9 or 10. The owners took a share for everything. They took a share for the capstan, a share for the little boat — yes, they took a share for every bloomin' thing! Then all the other stuff, the pots and pans and grub, that would come out of the expenses. That was another thing they worked. Mind you, the skippers didn't do too bad. They had a little touch orf the butcher and a little touch orf the grocer, see. No, they didn't do too bad. And they used t' git a backhander orf the coal. Oh, they got their perks all right. But the share system itself was a way o' diddlin' the blokes, that it was. It started so many years ago; it'd been the custom ever since there was a fish trade. It'd never altered. Course, afore the First War the driver had his own little racket. He was allowed the herrins what they called scummers. Yeah, scummers. When the crew were haulin' the nets the driver used t' stand there with a long didall an' catch the herrins what weren't meshed properly. See, they'd fall out an' he'd whip 'em aboard an' put 'em in a basket for himself. Praps he'd git a couple o' baskets full that way and they were all good, big herrin'. He had t' work at it, but that was his privilege, his perks. The crew got a little bit as well. There'd be a few herrin' what were gillded. They hadn't been shook out of the nets 'cause their gills were still caught in the mashes. Or there might be a few mackerel; not enough t' go up for sale, though, so the skipper would say t' the crew, "You can have them. Ol' so-and-so will buy 'em orf yuh."'

It fell to skippers to sort out the tangle of the shares system between members of the crew and for conscientious ones trying to be fair to all it could be a headache. One of these was Ned Mullender:

'When you carried an extra man your shares weren't so big. Like when I was in the *Hosanna (LT 167)*, we had a crew of 11 in her. Well, that meant you had t' come more ways, so that brought the share down t' £3-18-0d, somethin' like that. Usually the share used t' run out at four pound and half-a-crown a hundred. That was what that was always recognised as an' that was what the hawseman used t' git. He was full shareman, yuh see. Well, that took us more educated fellers what come inta the fishin' after our fathers' generation t' see that it should really ha' been £4-3-4d a hundred. See, we were gittin' done out o' 10d on the share. The owners used t' git about 60% o' the earnins and the crew about 40%. I forgit exac'ly what it was now; I couldn't tell yuh for sure. The whole thing used t' work in eighths, or half-quarters as they called 'em. You had different people on board an' they were all paid different. The skipper had a share and a quarter, and an extra half-quarter sometimes. See? Then you had the mate at a share an' a half-quarter. See? Then you had the hawseman and the driver at a full share. See? Then you come down to the three-quarter and half-quarter on the after deck an' the three-quarter and half-quarter on the fore deck (he was called the whaleman). Then you'd have the three-quarter shareman an' the stoker. Next was the cast-orf at a half and half-quarter, then there'd be the cook at a half-share. All o' them would git their shares according an' the owners would git the rest. I had it all writ down once when I used t' go t' sea.

'When I was in the *Impregnable (LT 1118)* I allus used t' give my cook an extra half-quarter, an extra eighth. See, he was gittin' a half an' half-quarter because he was a grown man, not a young boy, but I used t' make him up to the same as the three-quarter shareman. That was worth it. Cor, he was a cook an' a half! I mean, he could dish up anything. He never bought no bread neither; he allus used t' make his own bread. An' when you went down below t' have tea there was allus somethin' like buns or a bit o' cake. You know, there was allus somethin' to eat with your mug o' tea. Nine times out o' ten you got a three-course dinner in the middle o' the day. See? And he'd make it out o' nothin'. Well, not out o' nothin' exac'ly (don't git me wrong), but he used t' do it cheap. Well, that was good because the owners were always on at yuh about keepin' yuh food expenses down. They didn't like yuh spendin' too much. When we were round there at the Westward our gaffer was round there as well, a-sellin' the fish. Well, every now an' agin he used t' git Bill t' make him a couple o' loaves o' bread. Well, I didn't want t' lose a man like that, did I? I mean, I wasn't goin' t' lose a man what could feed the boat for about 12 bob a week per head. You know, 12 bob a man. That I wasn't, specially when another boat would be about 14 or 15 shillins. And another thing. Say you got a good stoker, like one chap I had along o' me. He could do anything on board a boat, mend nets, the lot. Well, you want t' keep people like that, so I got him an extra half-quarter. I used t' tell my guvnors an' explain to 'em an' they'd let me do it. "Well, you're the one who's responsible," they used t' say.'

Hard earned money is also hard lost money and Herbert Doy adds a postscript by explaining what happened to driftermen who, through illness or accident, suddenly became unable to work:

'When I was in the *Cyclamen (LT 1136)* I'd only bin on the hoom fishin' about three weeks and I poisoned both hands. That was orf here, orf Low'stoft. I got pricked by scads. Both hands! Yes, I had both hands messed up that hoom fishin'. Poor ol' Davy Spindler, he took my place. He went hawseman. Do yuh know, that was the first hoom fishin' I'd done in her an' I'd bin in her nine year! We'd bin trawlin' all the rest o' the time. See, she was a drifter-trawler. Anyway, I got messed up an' I had t' come ashore. That'd be about 1929. The owner, Porky Tripp o' Kessingland, he took me t' hospital. I couldn't do nothin', yuh see, I had both hands punctured, an' I had t' pay this man 10 bob a night t' take my place. Well, that went on for about a fortnight, so I said t' Porky Tripp, I said, "Porky, I'm goin' t' pack it in. I can't afford t' go on. "See, I was payin' 10 bob a night t' the man what took my place on the boat. That was comin' out o' my money. You were on a share o' the earnins, but what you paid a night-man had t' come out o' that. That was deducted from your earnins at the end o' the voyage.

'Like I said, poor ol' Davy Spindler was the man what took my place. He originated from Kessingland an' all. There was two or three of 'em about the market in them days who used t' go as night-men — ol' Joe Utting an' ol' George Sharman. They'd fill in at short notice, yuh see, an' they'd git more out o' doin' that than you would workin' full time. See, they'd git the 10 bob for nightin' an' a share o' what the catch made. Say they come in with a shot o' herrin', well they'd git a share o' that. They'd git the proper share. Say a night-man went as cast-orf, he'd git the cast-orf's share. That was all worked out each trip an' that's how they got the money. That was the only thing some of 'em used t' do, go night-man. They'd wait foʀ it, wait for a night-man's job. Yes, they got paid better'n you did. See, they got their 10 bob and a share o' the trip. Oh yes, a night-man could do better'n what the regular people did.'

Picking mackerel and scads out of the drift nets, a task heartily disliked by the fishermen, especially with the latter species which were painful.

CHAPTER SEVEN

Life on Board

'The Lord sent the grub
And the Devil sent the cooks.'
(Traditional East Anglian saying)

Talk to any drifterman about domestic life on board and he will wax eloquent about the shortcomings in food, accommodation and sanitary arrangements. Then he'll finish up by telling you that it wasn't such a bad life after all. The apparent contradiction needn't trouble us unduly. One has to remember that in the heyday of the steam drifter, before and after the First World War, standards of hygiene and comfort in most homes fell well below today's standards. Things taken for granted now were often far from general then, a good example being the presence of flush toilets. As well as the different level of expectation, one has also to take into account the capacity of human remembrance to see things from afar as being better than they really were. And I have detected frequently, among the men I talk to, the desire to compensate for a view of their working lives that they feel may have been exaggerated.

None of them need have any such misgivings for without any colouring life on the drifters was both tough and dangerous. The cruel sea image may have been excessively dwelt on by writers but it was real enough. The risk was always there, whether they were riding out a storm on the mighty Atlantic swell round at Westward or dodging a sudden squall in the close, treacherous reaches of the North Sea. It was at times like these when the weight of fish in the hold could result in a boat foundering, or when shipping a sea broadside on could wreck the wheelhouse, or sweep men overboard. The toll taken in human lives was heavy and the drifters lost around our coasts still occupy a prominent place in local lore. Names like the *Shorebreeze (LT 1149)*, the *Playmates (LT 180)* and the *Girl Norah (LT 1137)* will be a long time dying.

Even without the worst excesses of the weather, there was plenty for the drifterman to contend with. Working through the night on slippery, rolling decks, for instance, with only the odd break to spell round and swallow a welcome mug of hot tea. Added to this was the corrosive effect of salt water on the hands and other exposed parts of the body, while rest from fatigue could only be had in the cabin, where the bunk spaces were cramped. Mind you, most driftermen will say quite readily that life on board was made tolerable by the fact that individual trips usually only lasted one or two nights, and therefore the boats were always in and out of port. During the autumn fishing this meant nights at home, and even on the voyages to distant places the entertainment available in any particular harbour (customarily of a liquid nature) was usually enjoyed to the full.

There needed to be some small attractions to compensate not only for the cramped conditions on board, but also for what today would be termed "inconvenient hours". Then there was the safety factor. It would be wrong to suggest that there were no precautions taken on the drifters where life and limb were concerned, but a lot of what went unchallenged 50 years ago certainly wouldn't pass unnoticed today. The engine room is

THE DRIFTERMEN

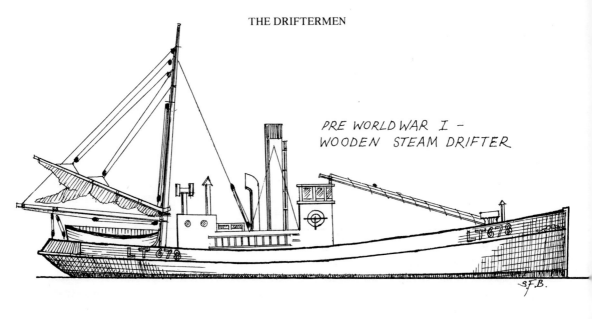

PRE WORLD WAR I - WOODEN STEAM DRIFTER

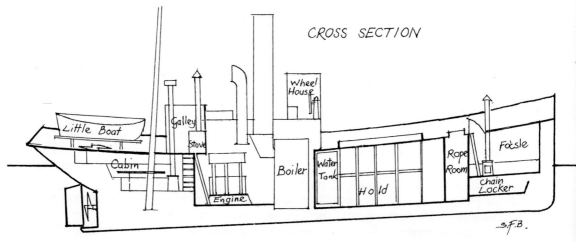

CROSS SECTION

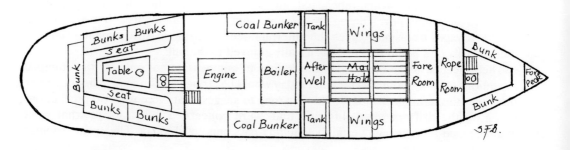

BELOW DECKS PLAN

LIFE ON BOARD

one place where tighter regulations now have to be observed and where there has to be much more guarding of moving machinery parts. An added hazard in this area at one time was the safe storage of drums of calcium carbide, carried on the boats to provide acetylene lighting. Explosions on board the steam trawlers *King Harald (GY 1097)* and *Betty Johnson* in 1917 and 1922 respectively resulted in an official enquiry. The Lowestoft Vessel Owners Association consequently decided to recommend to its members that the chief engineer should be personally responsible for seeing that drums of carbide were safely stowed away before the boat left her moorings and before the engines were started.

In the event of crew members being hurt while they were on board ship, there was a medical kit available to treat minor injuries. Its contents were laid down by the Board of Trade under the provisions of Section 200 of the Merchant Shipping Act of 1894, and they varied according to the type of fishing vessel. Steam drifters carried a basic first aid case, containing various dressings, scissors, forceps, eye drops, iodine and sal volatile. Then there were such various exotic additions as cramp mixture, diarrhoea mixture, stomach mixture and cough mixture. In the event of serious injury an immediate return to port was necessary, yet it was not unknown for some of the harder skippers to refuse to do this because a broken trip meant lost earnings. So there were occasions when badly hurt fishermen had to wait long hours for much needed treatment. They were a tough breed, and their attitude to life and death is best summed up by the number of them that couldn't even swim a stroke. What was the point, they ask, when if you went overboard in an oily frock and leather crotch boots you would sink like a stone anyway?

Fatalism or reality — which is it? Working on the drifters certainly had a lot to do with reality; the sting of the salt spray, the tug of the lint on calloused fingers and the roll of the deck underfoot. There was also the absence of material comforts, the little extras with which man always seeks to cosset himself. Life on the boats was basic. No hot and cold water at the turn of a tap and no proper lavatories either. The only area where the driftermen had any advantage at all was in the matter of food. Crews usually ate well, for although the owners were always eager to keep down expenses they did realise the importance of good rations and the effect of an adequate diet on the mood and performance of the men in their employment. Talking to men like Billy Thorpe the subject of rations always looms large:

'They were tough times. There was one bloke, he was seasick all trip and no one didn't see after him. He got in the small boat and he died of exposure in the end. The skipper got charged through that. Negligence, they called it. Well, someone shoulda seen arter him. They just let him lay there 'cause he was seasick. O' course, it turned out he was really ill and in the end he was vomitin' blood too. But that used t' be like that. As long as you could stand you were all right. When you couldn't stand on yuh feet the skipper would tell yuh t' go down below. That took me two or three year afore I got used t' bein' seasick. I never used t' have a dinner or a tea the first day out — never did for all the time I went t' sea. I knew if I had my dinner or my tea I was goin' t' sick it up. I wun't actually seasick, but what we call qualmy. When you were like that you never knew whether you wanted t' be sick or go t' the toilet, or where you wanted t' go.

'We only used t' take enough water for drinkin' on board the drifters, thass all. You didn't even have enough t' wash. In a steamboat you could put a bucket over the water which had been through the engine. That was a bit warm and you could wash in that. When you first went you started orf as cook. They used t' give yuh a big bit o' meat an' tell yuh t' put it in the oven. Then you had t' put so much water in an' cut an onion over the top. Then you'd peel yuh potatoes an' do yuh greens an' put them in. Then you had t' make yuh light duff. Now there's an art in that. You're gotta roll it up an' finish it properly. If you didn't roll it up an' finish it orf properly the dumplins would come up as what we called split-arses. They were split because the water'd got in. Well, you can picture a young boy straight out o' school havin' t' do all that. He needed some help.

'Once when I was for'ad in a boat called the *Lanner (LT 1176)* — an' you had a job gittin' for'ad on her because she was so narrow — she took a sea and I picked myself up right aft. Right aft, there you are. And I've still got a big scar on my head where I got hit. I had my sou' wester on an' that went right through. Some o' the blokes even got washed down the hold. There used t' be some rum things happen. I can remember haulin' one night when I was cast-orf aboard the *Fisher Boy (LT 334)*, and when there was a breeze the mate would often come for'ad with yuh t' git the seizins orf — specially if you was haulin' mackerel nets. Well, the rope flew out o' the molgogger an' caught the mate. I can picture him now right up in the air, an' he went inta the sea. Now we was all singin' out 'cause we could see him layin' on top o' the water with his oilskin keepin' him afloat. We got alongside him with a bit o' luck, but we couldn't git hold of him straight away. He went right under the ship more or less an' we got him the other side. He was a chap called Fred Hunt. I'll never forgit that. Oh yes, the ol' rope could fly out o' the molgogger!

'We used t' have wirelesses on board, but they were only receivin' sets at first. I can always remember — they never used t' work prop'ly. You used t' hafta bang them. You allus had a bit o' wood lay alongside of 'em t' tap 'em with every now and then. I can remember ours in the *Ocean Sunlight (YH 28)*, that didn't go for weeks. And I can remember our mizzen boom breakin' an' that come down solid on the galley top — an' after that the wireless worked! One ol' boy who was skipper in the *Silver Dawn (LT 194)*, he used t' come out with a lot o' comical sayins. We come alongside him once an' he say t' us, "Cor, they give out a bloomin' anti-cyclone! I'm goin' back in." He dint know what the word anti meant. He thought that was goin' t' blow three gales together! Another one o' his was when the French-letter boots first come out (they were made o' rubber, see, not leather). One bloke slipped over on the deck. "Well," he say, "you will wear these here bloomin' rubber heel irons an' they're what make yuh slip!" He was a rum ol' boy, he was.'

Jack Rose was a member of the last generation born into the era of the real steam drifter but well remembers some of the primitive domestic arrangements:

'The old ways started t' change in the fifties. The boats as well. Up to the fifties there was no toilets or sanitation aboard these boats at all. Some of 'em had a tub, but there was several drifter-trawlers I went on never had nothin'. Fine weather you'd go over the side, over the rail; bad weather you'd go down the engine-room. You'd git a shovelful o' coal, squat down over the shovelful and hull it straight inta the furnace. That was like steamin' hoom — sometimes the ol' cook or the ol' engineer wouldn't let yuh have no water t' wash with. So you'd git yuh water in the draw-bucket from over the side and take it down the foc'sle. Then you'd git a shackle or suffin, drop it in the fire till it got red hot an' then drop it in your bucket o' water. An' that was the only hot water you got.

LIFE ON BOARD

'Years ago the ol' skippers used t' number the medicine bottles 'cause half the blokes couldn't read or write. They used t' number the bottles an' someone would go down an' say t' the skipper, "I're got a touch o' the diarrhoea" or suffin like that. "Think of a number," the ol' skipper would say. "Number 6," say the bloke, so the skipper would git out number 6 an' give him some. That could be liniment or anything! That dint matter; they'd give yuh a spoonful of it. When I first went t' sea I had toothache. And do you think I could git rid of it! Do yuh know what I was doin' most o' the time? — gittin' brown paper, puttin' pepper in it, puttin' it in the oven, bringin' it out and layin' my face on it. That was the way I got heat treatment! They wouldn't bring me in and that was a hollow tooth. I was stickin' bacca in it and I'd chew a bit o' bread an' put that in, but none o' that worked.

'Do yuh know, I have bin 18 hours haulin' nets. You're on the go all the while and you don't stop for grub like the Scotties do. You'd eat when your work was done. And I've gone down an' had my dinner on the table there an' my face has gone in it an' I've gone t' sleep in the hot gravy. And that's no lie. That happened on the ol' *Present Help (LT 1120)*. Then, when you had all yuh nets in, you'd hafta pull 'em all up on deck agin when you got inta harbour. That was t' clean 'em. Then you'd pull 'em down agin the next mornin' afore you went out. When you were at Shields that weren't too good. You got jellyfish there. The water used t' sting yuh, never mind the jellyfishes! I've seen the blokes go down the engine-room an' wash theirselves in paraffin an' open the furnace door an' stand there t' burn it out. And I've seen 'em put buckets with eyeholes in over their heads when they're scuddin' out. Even the beatsters who used t' mend the nets used t' git affected by jellies, what with itchin' and so forth.'

What you might call a saga of discomfort, that story of Jack's. But there was the solace offered by good grub. Jumbo Fiske's is the authoritative voice:

'That was good food aboard the boats. Yes, good food. Always good food aboard the boats. There weren't many people lived ashore like they lived aboard the boats. You used t' git a good ol' beef puddin', or rhubub duff, or suet duff. Plenty o' things like that. Plum duff. Plum duff o' Sundays. Oh yes, that was good food aboard the boats. Some people said it weren't no good, but there weren't many people lived as well ashore as what they did aboard a boat. There was always a five pound cookin' o' meat; a good dinner an' that, yuh know. Then there'd be a bit o' ham; you'd have ham an' that for tea. Course, when you were driftin' you had t' git yuh meals when yuh could. When you had a young boy as cook there'd always be someone t' put him right. Yes, there would. If he wanted help, the men would help him. If you had a good cook you could live well, that you could. The grub was there t' cook if you had a cook who could cook it. That was why a lot o' the cooks were full-grown men on a three-quarter share. They'd stay as cook on that rate o' pay.

'As regards medicine, you always had corf mixture on board, yuh know. That was well equipped with that sort o' thing: corf mixtures, diarrhoea mixtures. And there was hell an' all bandages if you wanted t' bandage anybody up. Yes, we had a good first aid kit, that we did. Them boats were well equipped, yuh know I don't spose the early boats were too good, but when I went they weren't bad. When we were in them *Hides* (*Margaret Hide* and *Sarah Hide* mentioned earlier) we had a good kit. That we did. A gret big box. We had the ol' sea biscuits too. Christ, they were hard! But do yuh know what? If you had one o' them an' soaked it, then toasted it, that was really lovely. We never had that many accidents on board, but you'd git one every so often. When I was in the *Olivae (LT 1297)* the cast-orf went over the side. You know how the boat go up and down on the sea? Well, he was standin' aft side o' the mole-jenny, where the rope run through, an' the rope flew out an' knocked him clean over the side. We got him all right, though. You'd git the odd bloke fall down the hold as well. Then I expect you're heard about how chaps have gone round the capstan with the rope, when they were haulin' the nets, though I han't never seen that myself. When anybody in the crew was a little bit queer I used t' give him a good ol' drop o' hot rum. I used t' keep a bottle in my bunk. There weren't no particular rule about takin' drink on board an' when you went away on a voyage you allus had a bottle o' whisky. After we got out o' harbour I used t' say, "Give the boys all a dram." Then the mate used t' go round an' give 'em all a tot. I allus made that a practice.

'When you were on a long steam, like down t' Shetland, you used t' play cards. You'd have a game o' crib like, or a game o' phat. That'd be the tens and fives, the phat would, not the nines and fives. Sometimes the blokes used t' read. You know, have a read an' then turn in till that was their turn t' go on watch. That weren't too bad on board really. I dare say that musta been crowded in them little ol' early boats, but I never did go in one o' them. Some of 'em dint even use t' have a table; yuh used t' eat yuh food on the floor! Some o' the ol' boys used t' tell yuh that. You'd git down t' the platter, they said, an' eat on yuh knees. Mind yuh, a lot o' the boats in my time dint have toilets on board. You used t' cut a barrel in half, or an oildrum, put some rubber round the top an' sit on that. Then you'd chuck it overboard when it was full. We had toilets in the *Hides*, though. Well, we had a seat with a pan underneath. At least that was somethin' where you could git orf the deck, otherwise you had t' sit on deck an' do it. When you were in harbour you used t' do the wheelhouse windows over with some whitewash stuff an' go in the wheelhouse. You know, put yuh tub in there. Some of 'em even used t' go down the engine-room an' squat over a shovelful o' coal.

'You used t' live fairly close together, but, touch wood, I allus had a good ol' happy crew. Yes, we were allus a good ol' happy flock. And when we made up we used t' have a good day. Yes, when we finished a voyage we used t' go on a bloody bender. That we did! There used t' be a real beano that day. I never had many fresh chaps in my crew. We were always together, the same chaps. That's the best way t' be. You look after them an' they look after you. Some o' the ol' skippers, though, they used t' be lazy ol' devils. Lazy as anything. They wouldn't do a thing. No! You had t' call 'em out o' the bunk when you'd done haulin' an' all that sort o' thing. They never had t' do wi' me, thank the Lord! People used t' say, "Cor, you don't half earn some money!" Well, you could do if you looked out for what you were doin'. That didn't do t' be asleep. Some of 'em used t' come aboard drunk, yuh know, an' they'd be asleep till the crew had done haulin' the first shot. Well, you dint earn nothin' that way.'

As Jumbo Fiske earned more money catching herring than any other skipper of the steam era, his remarks can be regarded with authority, but Ernie Armes remembers some time honoured remedies that may not have been in Jumbo's excellent first aid kit:

'Everything what fishermen tell yuh about how bad things are today, the poor ol' boys years ago were a bloomin' sight worse orf! Nowadays they're got morphine an' dickens know what else t' kill pain, but in the ol' days there wun't nothin'. The poor ol' boys used t' have bandages round their wrists soaked in paraffin, red flannel bandages, t' stop the chafin' from their oilies. Then they used t' git stung by jellyfish, particularly up in the northern part of the North Sea. Cor, there were some bloody gret jellyfish up there. They'd git stung all right! Then they'd have salt water boils round their necks an' all round their wrists where the oily frock caught 'em an' chafed 'em. Cor, gret big ol' boils they were! Nowadays, if they have a boil on the tip of a finger they go up t' the doctor an' git a month orf on the club! Thass enough t' make yuh sick.

'Mrs. Love's Ointment, that was the fishermen's friend at one time. That it was — Mrs. Love's Ointment. Ol' Mother Love, she lived down Denmark Road an' she used t' make the stuff in her back room. There used t' be one or two little grocery shops say, "Mrs. Love's Ointment sold here." That used t' come in little wooden boxes an' my father swore blind by that. That an' Friar's Balsam. Yeah, Friar's Balsam cured evrything. If they'd had a bloomin' toe ache or a cut throat, I believe they'd have took Friar's Balsam! You'd git a spoonful o' sugar an' then put the Friar's Balsam on, three or four drops, an' swallow it. They used t' take Sloane's Liniment on sugar as well. For sore throats, that was. Another one was Carlton's Dutch Drops. Same thing. You'd put it on sugar an' take it. All the fishermen used t' swear by it. Just like I said a bit earlier about Mrs. Love's Ointment, everyone used that. If they got pricked by a scad or a gurnet, they'd use hot vinegar first (they reckoned that used t' kill the poison) an' then bandage it up with Mrs. Love's Ointment. There weren't no gittin' the doctor in them days. If a man got hurt he had t' put up with it. The skipper wasn't goin' t' break his fishin' trip because a bloke felt ill — not unless he broke his neck. Then they might bring him in. There was a lot o' accidents on board in them days, but nothin' must stop the fishin'. That had t' go on. If you were queer, you were queer; you still had t' do yuh work.

'I went out on the boats more'n once when I was a boy. You'd go out for a pleasure trip hoom fishin' time. I went with my cousin Jimmy once in the *Pride (LT 471)*; that was his father's boat. We come hoom from school the Friday mornin', dint go back in the afternoon an' rushed orf down the market. "Come on," my uncle say t' Jimmy an' me, "jump aboard." Well, I was seasick within the hour, an' was I bad! Do yuh know what? — they hulled me in the little boat. Yes, I hetta lay there. They covered me up with a bit of ol' barmskin. They wouldn't let yuh go down the cabin an' be sick — not till that got dark anyway, an' then they hulled yuh down. They were little ol' cabins in them days, yuh know. We had t' lay on the after-locker an' we were both bad, both us boys. And I got covered with bug bites! I was from my neck down with 'em. When I got hoom my mother say, "Take them clo's orf!" So I hetta go in the kitchen an' have a good wash down an' put some clean clo's on. "You aren't goin' there no more!" say my mother. "I shan't allow you t' go no more." But I did go agin. Course I did.'

The last remark says a lot about a young lad's outlook, but going on a pleasure trip (pleasure?) and going because you had to earn a living were two completely different things. Frank Fisk remembers that his trips were out of necessity:

'After I got married I did nearly all driftin'. You had no facilities, yuh know, as regards washin' — only a bucket. There weren't no toilets in them days neither, only a round tub which you all had t' use. If you wanted t' go ashore when you were away on a voyage, you all had t' wait for each other till you'd done washin' in the buckets, in the pails. When you went t' bed you used t' lay on a bag o' straw. Yes, that you did. When I went away on a voyage I used t' hafta go to a farmer and git some straw t' make a bed from. There were bunks in the cabin an' you used t' put a straw bed in an' lay on that. You never took yuh clo's orf, yuh know. I used t' turn in just as I am now, sittin' here talkin' t' you, an' pull a rug over me.

'You always ate well on the boats. The stoker used t' clean all the fish for breakfast, all the herrin', an' he used t' wash 'em in a pail. Then the cook used t' fry 'em up. You always used pails for different jobs 'cause there weren't nothin' else. If you wanted t' git some water from over the side you used a bucket with a rope on, but you'd never twist that rope round yuh hand because if you did that might pull yuh over. You used t' have it so if that did try t' take yuh over you could let go. You know, you could just let go. No, you never used t' wind it round yuh wrist because that'd pull yuh over as you went along. Sometimes, when there was a bit of a breeze, a man would git knocked overboard. That used t' happen t' the cast-orf sometimes when he was takin' the seizins orf the rope. That rope would jump up out o' the mole-jenny an' knock him over the side. Oh yeah, there's bin many a man lorst like that, specially if there was a bit of a swell on. I mean, you went t' sea in all weathers. Oh yeah. That'd got t' blow hellishly hard t' stop yuh from goin' out.

'Sometimes that'd be mountains high! When you went round t' Westward that was like that. When we went round t' Newlyn the sea would go right up high. Well o' course you'd go up with it an' then you'd come down. Sometimes you used t' wonder just where you were goin'. You allus kept the boat inta the wind. That'd knock along for evermore if you kept inta the wind with the sail up. We used t' be out for a week at a time in bad weather, a week at a time, an' we didn't fish at all because that was too rough. All you did was keep dodgin'. You put yuh mizzen sail up, set that dead in the middle, an' then you went in the wheelhouse an' set the wheel. She used t' be so steady that you could go t' sleep an' let her dodge along. Yes, you could dodge about for hours an' you'd be all right. The only thing you mustn't do was let her go orf across the wind or else you would have a bit of a flap on.

'I can describe when we were round there once at Newlyn an' we all had t' come in because o' the weather. That was as fine as silk at dinnertime, about 12 or 1 o'clock time, an' then that come on a gale about two. We had t' batten down. Yes, we battened down everywhere an' then all the wheelhouse windows got knocked out. We managed t' git inta Newlyn all right, but nearly every drifter lorst a mizzen mast or a little boat or suffin. That was a proper rough day that was. One o' the boats — that mighta been the *Welcome Friend (LT 375)* — actually lorst her wheelhouse. That was washed away by a sea an' she hent got nothin', only a plain deck. Her mast was gone as well an' the bulwarks were all broke. I don't know how she managed t' git inta Newlyn, but there was a crowd o' people t' see her come in. Yes, we used t' git some rough weather round there — orf the Bishop Rock an' down the Lizard an' the Longships. That was marvellous how them little ol' drifters used t' stick it.

'When we went away on a voyage we all used t' have a little ditty-box t' keep medicines in an' writing paper an' what little money you had. Yes, we used t' call that a ditty-box. We had one each. That was just a little wooden box; I used t' bring mine hoom at the end o' the voyage. You used t' have embrocations an' things like that in it. Mrs. Love's Ointment as well. We used t' take that for the hands. Up at Shields you used t' git these jellyfish. Cor, they were murder! They used t' sting yuh, an' I've even known the blokes t' put reed bags over their hids an' cut holes in for the eyes, when they were haulin' in the nets. They did that 'cause the jellies stung so. After you'd done haulin' you used t' wash in the warm water what come out o' the engine room. You used t' put yuh bucket over the side, collect it an' wash yuh hands in it so that'd take the sting away.'

Despite the hazards and hardships there was often a comical side to life at sea and George Stock reveals a glimpse or two of it when he discusses his early experiences as cook:

'I come orf the smacks about 1920, 1922, an' I went on the drifters. That was a bad time round about then. A lot o' the men were workin' on the sea wall, drawin' relief an' that sort o' business. I went cook in the little ol' *Welcome Home (LT 402)*; she didn't even have a table down the cabin. We went down t' Shields first of all, then we come t' Low'stoft for the hoom fishin'. We earnt her number. Her number was 402 an' that was how much money we earnt for about 16 or 17 weeks — £402. We paid orf in debt, but I'd bin gittin' 7/6d a week allotment an' we all had lines as well. You know, hand-lines. We used t' git coalies an' cod on them, an' when we went t' places like Scarborough we'd git a good price for 'em. We used t' sleep on straw beds. Donkeys' breakfasts we called 'em. Yeah, we used t' buy 'em at Yarmouth Stores for half-a-crown.

'When I was in that *Welcome Home* there was a breeze one time an' that blew so hard we were out four or five nights altogether. You couldn't git t' work an' that upset the beef kettle so it fell orf the stove an' broke. Well, we used t' have our fat for fryin' come in buckets, with a lift-up lid, an' that's how I cooked the meals for four days — in this here fat pail. When we come in, the *Nevertheless (LT 1148)* — I think that was her name — was ashore in the harbour mouth so they warned us orf t' go t' Yarmouth. But the skipper, ol' Barney Weevers, he say, "We're goin' in whatever happen!" So we all got down below an' shut the galley door. Well, we got in all right, but we took all our rail orf down one side an' we had t' put the boat straight on the harbour beach.

'When you were cook you used t' make yuh light duff. The men used t' like that. There was eggs in it. O' course, you had t' be clean an' then everything was all right. Yes, you'd make yuh light duff an' then sometimes you'd make the suet duffs, which was with the ordinary block suet. That used t' come from the butcher's an' that was nice to eat as it was. You'd grate it up an' make yuh duff that way. When it was done you weren't allowed t' cut it with a knife t' see if it was ready. All you would do was put a fork down on it an' you'd see it all glisten. After the boys had had it with gravy an' meat, they used t' have it with treacle or jam. Oh, that was nice — not at all like this here packet suet you git today. Cor, I could fancy one o' them now; my mother used t' make 'em a lot.

'Another thing you used t' have were the ol' sea biscuits. Cooper's rusks we used t' call 'em. Hard as iron they were; you could nearly throw 'em through the wall! The longer you had 'em the more an' more full they used t' git with them little gold an' silver weevils. They'd be through an' through these biscuits, but we never paid no regard. We used t' soak 'em an' then put 'em in the oven. That *Welcome Home* I was on, she didn't have no table down the cabin — I believe I told yuh. Well, I remember a funny thing happenin' on her. I always used t' keep a nice scrubbed sack, cut down an' opened out, which I'd put down on the floor to stop stuff slidin' around. One o' the crew, the mate in fact, was an ol' country chap; a terrific big bloke he was. Well, one day I got the dinner out an' put the beef kettle down on the floor an' sung out, "Dinner-oh!" This ol' boy I told yuh about, he musta bin in a dream. He come flyin' out o' his bunk an' his feet went right in the beef kettle! That was scaldin' hot. Cor, you ought to have heard the language!

'That was like when I was along o' my brother in the *Castlebay (LT 1295)*. He was skipper an' the two of us were always after rats. We used t' have a boot-jack t' chase 'em with. You know, that was a piece o' wood with a block underneath and a notched end; you'd pull yuh boots orf with that. Anyway, we saw this big ol' rat goin' round the cabin so we shut the door. My brother, he got up out o' his bunk an' he say, "Leave it t' me." But I was just then swipin' at it; I didn't know he was out o' his bunk. I went t' hit this rat an' I hit his big toe instead. He chased me round the cabin. "I'll chop your b----- hid orf!" he say. Things like that, you never forgit 'em, yuh know.'

A man on the receiving end of a death-blow intended for a rat wouldn't forget it in a hurry but rats as shipmates were taken for granted. Horace Thrower can also remember some fun and games with them:

'You used t' git a lot o' rats on board the drifters. They'd be in the wings an' down in the fore-room where you kept yuh spare gear. I've even known 'em in the little boat, even though it was all covered up. Oh yes, I've seen rats in there. When I was in the *Tritonia (LT 188)* we once had a rat git behind the steam pipe what run along underneath the rail t' the capstan. After we shot that night the cat we had aboard kept goin' inta the kid. You know, that kept marchin' up an' down, a-lookin' at this here pipe. Well, we couldn't make out what was wrong at first. Then Doff Muttitt, the skipper, say, "I don't know. I think there's somethin' behind that pipe." See, that was all lagged. Well, I'll tell yuh what we done. We thought that'd be a rat most likely so we put the deck hose in one end an' we turned it on full force. Out come the rat the other end an' the cat grabbed hold of it! That knew it was there, but couldn't git at it till we turned the water on.

'That was a rum little ol' cat. We once lorst it overboard. Yes, that fell over the side. That would walk along the rail an' we'd got the sail up there once on the foremast. We never worked it a lot; that was only a small one. But the cat must have been comin' along the rail an' the sail must have started t' flap an' caught the cat an' threw it overboard. We went full astern an' then we had t' try an' git the damn thing on board. Do yuh know what we did? We got a quarter-cran basket an' trailed it along the side an' got it in that. I'll allus remember that. Emergency — cat overboard! We knew there was somethin' wrong because we were down below when we felt her go astern. When we come up on deck t' have a look, the skipper told us what it was. I thought I'd seen that cat go over several times before that; the little ol' devil would walk along the rail in bad weather. Anyway, the sail did catch it this time an' over it went.

'There used t' be accidents t' the crew as well. One thing you had t' watch was the capstan when you were haulin' the nets. Praps the cast-orf would git his oily wrapped round the rope an' that'd pull him round the capstan. I never heard o' no one gittin' killed that way 'cause you could soon stop the capstan. There was a thing on the top what you could shut if orf with. But there have been accidents like that, specially when they were takin' seizins orf the rope. Praps one wouldn't come undone an' that'd go round the capstan, see. Well, the cast-orf might try an' git it orf without stoppin' the capstan an' then

LIFE ON BOARD

praps his oily would catch in it. Another thing you had t' watch was when you were shootin'. You used t' pull the nets up from the hold over a roller an' if they come foul that could be awkward. We lorst a chap round at Plymouth through that. The net got foul o' the roller, pulled the roller up, swung round an' caught one o' the chaps what was shootin' the nets an' threw him overboard. That was early one mornin', about the second time we were shootin' (we dint git nothin' the first time). We tried t' git him. We threw a lifebelt at him an' got him alongside the ship, and' we were just goin' t' put a rope round him, underneath his arms, t' pull him aboard, when down he went. We never saw him no more. A chap Utting his name was; come from Kessingland. Whether he hit his head a-goin' over an' stunned himself, I don't know, but I should say that he did. We all had t' go t' the Custom House an' make a statement; all separate, you know. That was a rum business.

'But o' course life was rough on board. There was no comfort. I mean, you used t' have straw mattresses on yuh beds. You used t' fill 'em up yourself an' take 'em aboard. You know, you'd go an' see an ol' farmer an' git the straw. The latter part o' the time they used t' sell 'em in some o' the stores, but when I first started you had t' fill yuh own up. Then after you'd had 'em praps about a year they would all git smalled up inside an' the straw used t' poke through the hessian. Yeah. We used t' chuck 'em overboard then, or burn 'em. You could stick 'em in the furnace if you wanted, but you usually used t' wait till you'd finished the voyage. Course, you never knew when you were goin' t' finish or when you weren't. They never used t' tell yuh nothin'. I mean, like at the end o' the hoom fishin', praps gittin' towards Christmas, there'd still be a few herrin' about so you'd keep goin' after 'em. Then all at once you'd come in one day and they'd say, "Right, you'd better finish."

'Another thing was you didn't have no toilets on the early boats. You just had an ol' tub, one of the ol' Dutch bowls cut in half, an' you sat on that. That'd be up on the deck somewhere. In bad weather you used t' go down the engine-room an' squat over a shovel, then hull it inta the furnace. I've done that more'n once. You can't go an' sit on the deck when there's bad weather. I think some of 'em used t' do it more down the engine-room than they did on deck! Whether that was fine or not! The engineers used t' do it theirselves, so they didn't mind. I mean, that all used t' burn away. The blokes used t' come down the engine-room t' have a wash as well. There used t' be a big platform at the back o' the engine an' they used t' stand on that, where all the pumps were. That was just afore yuh got t' the cabin. You'd got a bulkhead there at the back o' the engine an' then the other side o' that was the cabin. There used t' be this nice big platform up that back end an' you used t' keep that nice an' clean for the blokes t' wash on. You know, you'd scrub it two or three times a week. You used t' git the water out o' the geyser, what they called the geyser. That come orf the boiler an' you used t' git steamin' hot water out o' that. You just used t' put the ol' bucket underneath an' turn the cap on. Cor, that used t' give yuh a lovely wash.'

It says a lot when a bucket of hot water is talked about almost in terms of a luxury. But that was the life of a drifterman. No frills and no comforts. Herbert Doy rounds off this look at the fisherman's life off watch with some of his early experiences at sea just before, and during, the First World War:

'I started as cook on the boats afore I was 14. I was cook for nine people. That was in the ol' *Fame (YH 854)* along o' Ned Mullender. (Father of the man who figures in this book.) I used t' coil a whole fleet o' ropes with a blinkin' duck lamp an' when I got out o' the rope-room my nose used t' be all full o' smoke an' smuts. Soon as you'd done that you used t' go straight away aft an' git breakfast. The chief or the stoker used t' git the pan all ready for yuh an' clean the herrin'. The crew used t' eat about six or seven herrin' apiece an' you used t' hafta fry that lot for 'em after you'd coiled a whole fleet o' ropes. That was the worst job there was aboard a boat an' the lessest pay. A half-share, that's all a cook got then. You had t' learn how t' be a cook, yuh know. The first lot o' light duff I ever made weighed about a ton! That was like dumplins; you know, the ol' 20 minute swimmers. On Saturdays you sometimes used t' have a plum duff, specially if you were goin' t' lay in. Then suet duff an' treacle, that was another meal. All the old stuff what people won't even touch nowadays.

'I went round t' Brixham for my first voyage. That was a herrin' voyage, when ol' Oscar Pipes an' them had motor boats. See, all the best drifters had gone patrollin' or minesweepin', so that only left these old converted jobs what had had motors put in 'em. And o' course the ol' *Fame* what I was in, she'd been converted from a sailin' boat as well, only she had a small steam engine in an' one the ol' Woodbine funnels. She wouldn't even steam aginst the tide, she wouldn't. We come out o' the harbour along o' them little motor boats an' they all got out an' we got nowhere. We had t' turn back; we couldn't git out aginst the tide. (She only had an 8 h.p. engine.) Yes, they all come motorin' up past us, all these other little drifters; they come motorin' up past us as if we were standin' still. Ol' Ned, he got hold o' the whistle. He say, "There's one thing you hent got — one o' these!" An' he give the whistle a pull. He used t' do that when we went out. He couldn't even git in the wheelhouse, yuh know. He was too fat. Poor ol' Ted Dunn, what was mate, he was the one who used t' go in the wheelhouse.

'We never did no good round there. No, we never earnt our salt. We caught a few herrin', but about 50 or 60 cran was a big shot in her. See, she was a little ol' sailin' boat converted into a drifter. You'd only shoot about 60 nets in her she was so small. The livin' quarters were very small as well. There were three up for'ad in the foc'sle an' the rest were aft. I was aft, in the cabin, an' I had an ol' Dutch oven t' cook on, one o' the ol' upright ones. I was such a big man I couldn't even lift the beef-kettle on the stove. That's the truth — that was too high for me t' lift the beef-kettle on. As well as cookin', you had t' scrub the cabin out an' everything. Yes, you had t' keep everywhere clean because some o' them ol' boys were very fussy. That they were. After you'd scrubbed places out, they'd come an' have a look round. Some of 'em even used t' lay aginst the bread locker so you couldn't git a sea biscuit out when you were hungry. Mind yuh, they weren't a bad ol' lot really — unless you weren't no good yuhself an' then you'd go through it. That you would.

'Another ol' drifter I was on, the *Resolute (LT 960)*, she had a funnel fore side o' the wheelhouse. Ted Chilvers was the skipper. He was a biggish man an' he was very religious. He used t' go t' the Bethel an' he wouldn't fish after midnight Saturday, so we used t' knock orf Sundays along o' him. He used t' play a concertina an' sing hymns. He was a very interestin' man, he was. I was along o' him in sailin' boats, in drifters and in steam trawlers. One ol' boat was called the *Searchlight*. She was an ol' Scotch runtie what'd had a motor put in her. We were out in her once an' we got clean swept down t' Cromer Knoll. That was about 1917, I think. We'd got about 25 cran o' herrin' an' that come on a gale o' wind. The skipper pulled the wheel hard over an' knocked the fangs orf the propellor with the rudder.

Well, then they got the mast up for'ad an' runned a sail up. Down come the mast! Then we took a sea, which knocked part o' the wheelhouse away. We were three days like that. The *Glory (LT 1027)*, a smack, held us orf the sand. He got his tow warp on us an' held us orf the sand till the *Albatross*, a destroyer, come an' got hold of us. When he tried the first time, the warp got in his propellor so a bloke had t' go down an' clear that. Once that was done he pulled us up the roads. Yes, he pulled us up the roads about 14 mile an hour! That was the *Albatross*. He tew us inta Low'stoft. We were supposed t' be lorst that time. Someone said they'd even seen the skipper's bank book floatin' about. An' that weren't a joke either! My mother an' all them were down there when we come in. They'd give us up as lorst.

'Cor, that was a rum job there for a bit. We nailed canvas all round agin the mast. You know, where that'd pulled up. We never had no grub aboard. See, when you're driftin' out here, you're out one day an' back the next. Well, we were out three days an' that meant we didn't have no grub aboard. We scraped the lockers out an' made some pea soup with the pieces what were in the lockers. An' the gunboat, he chucked some food aboard for us. Yeah, the *Albatross*, he chucked some food aboard — corned beef an' biscuits an' that. An' that's what we had till we got hoom t' Low'stoft. I'll never forgit that happenin' as long as I live.

'There was quite a bit o' herrin' fishin' out here durin' the First War. I was in another boat as well, the *Ocean's Gift (LT 387)*. She was a little ol' Elliott pot job an' ol' Mimsh Beamish was skipper of her. Oh yes, we were out fishin', but we weren't allowed t' show a light. You used t' have a little oil light show through inta the rope-room so the cook could see what he was doin', but that was all. That was all you had, just a little glimmer. Course, that was very dangerous then. I mean, I saw U-boats an' there was one time when our skipper shoved a mine away with a boathook. At that time o' day we used t' hafta wait for an escort before we could come in. There used t' be patrol boats an' they used t' escort yuh hoom.

'With the driftin', you had definite limits t' fish in. An' the escorts used t' be with yuh all the time you were movin' about. When you wanted t' go hoom you used t' run a special flag up, an' then he'd come an' git yuh all together an' away you'd come. You used t' have certain times when you could all git together, like the break o' daylight, an' then in you'd come. If you'd got any herrin', that was. If you hadn't got no herrin', then you wouldn't come in. Sometimes you'd git ordered in. Oh yes, when there was U-boats about. Then, when you got t' the harbour mouth, you'd see the owner standin' on the pier head. "Thass all clear now!" he'd shout. So we didn't go in. Oh yeah, we had a lot o' that — "Thass all clear now!" They were a hungry lot, them owners. That they were!'

"Snotched" – ready for the pan.

A fisherman in his 'walking out' dress of striped trousers, serge jacket with velvet collar, best gansy and black silk wrapper.

CHAPTER EIGHT

Custom and Belief

*'Wash upon a sailing day
And you will wash your man away.'*
(Traditional)

All seamen are superstitious and the herring fishermen were no exception. The subject would make a story for a book in itself. Fishing grew up with strong Nonconformist links, but sometimes religious belief and primitive superstition become intertwined. Much of what was believed in (or paid lip service, anyway) had to do with sympathetic magic, where an action could assume good or evil properties simply by association. A fine example of this is the way that eggshells were once smashed after the boiled eggs themselves had been eaten because it was believed that, if left intact, they might provide evil spirits with craft in which to put to sea and harm the sailors. It's not asking anyone too much, even nowadays, to shed some of his or her sophistication and see the reasoning behind that particular practise. Nor should we be surprised at the mixture of Christianity and paganism that led some old skippers to throw a few coppers over the side of the boat before the first net was shot to 'buy' a catch from Almighty God. The old traditions were strong as Jack Rose makes clear:

'You used t' git some o' these skippers that were a bit blasphemous. They used t' swear at God if they didn't git no fish. Then there were some that used t' chuck their money over the side t' buy the herrin'. I've done that myself. Ol' Nim Manthorpe was a big man for that as well. We were all very superstitious. I still am, but I was brought up that way, yuh see. I mean, if you didn't have a rat on board, that was unlucky. Mind you, most o' the boats were full o' rats. I went on one up t' Peterhead an' we caught 32 rats on her by the time we got back to Yarmouth! O' course, she'd bin layin' up for months afore we took her out. I'll tell yuh another thing too. I once got swore at by Jackie Muttitt 'cause I crossed knives down on the galley table an' that meant trouble. And I'll tell yuh suffin else I done along o' him which made him git on to me: I stirred a mug o' tea with a knife handle 'cause I couldn't find a spoon. Cor, he dint half bang the table! He say, "Stir with a knife, stir up strife./Stir with a fork, stir up talk." So what could yuh do? And another thing — you must never lay a hatch cover upside down 'cause that meant turnin' the boat over. Oh, there were 101 superstitions. You weren't even allowed t' whistle 'cause that meant you were whistlin' up a gale o' wind.

'If you saw a nun, yuh know, the first thing you'd do was shout out, "Iron! Iron! Iron!" An' then you'd run an' touch iron. I've even sin some o' the old boys touch the blakeys (studs) in their boots. There used t' be a joke about that as well: why do nuns always go about in pairs? So that one nun see that the other nun don't git none! Yeah, that was a real ol' fisherman's sayin'. At Low'stoft the blokes dint like parsons goin' on board ship. The only one they'd let on was ol' Tupper-Carey. Yes, he was the only parson they'd let aboard down here. (The Rev. A.D. Tupper-Carey, Vicar of St. Margaret's Church, Lowestoft.) I suppose they classed him as one o' themselves because he got so popular with 'em. Oh, there were some funny beliefs at one time. Like the day yuh father went away t' sea, yuh mother would never do her washin' for the very simple reason that if she did she'd be washin' him away. I never knew my mother t' do her washin' on the day my father went t' sea. And that's why I'm superstitious today, yuh see — I've been brought up with all that sort o' thing.'

Billy Thorpe continues in very much the same vein:

'When you shot the nets, you always said "In the name o' the Lord." Yes, that you did. It was supposed t' remind you o' the time when the disciples caught all them fish in the Sea o' Galilee. Then there was what was called the King Herrin'. The ol' boys would suddenly pick this fish out o' the net an' look at it an' chuck it back inta the sea, hopin' that was goin' t' bring luck. When we were young chaps we used t' see the ol' men pick this fish out o' the mashes an' that took you a while t' find out what they were lookin' at. I believe it was something t' do with the freshness of it, or the bluish colour, or something in the tail. That was something they knew, anyway.

'On some o' the drifters the drivers would have a friend — a rat. Every night the ol' rat would come out an' see the driver. They even used t' feed 'em, some o' the drivers; I've heard 'em talking about it. But o' course some animals weren't very popular. Years ago, when we first had wirelesses, they used t' give out the stock market prices after the news. But before that could come on the skipper would jump an' say. "They aren't goin' t' talk about pigs an' cattle an' rabbits on board my ship!" You know, they used t' give out how much they'd made an' so forth, but the skipper would switch it orf. Pigs, pork, anything like that — you mustn't mention 'em. And another thing: say you'd bin washin' the decks down an' you laid your broom on top o' the nets; the skipper would be on to yuh like a flash. Yis, that he would. You must never lay your broom on top o' the nets because that was like sweepin' 'em away. And you weren't allowed t' whistle either. The skipper would say, "You'll hear enough o' that blowin' when there is a gale, without you whistlin' it up!"

'Oh, there were some funny ol' ways. That was like the clo's the fishermen wore; there were particular fashions. For a start, we all had high-heeled boots. I can always remember my sister (I'd bin at sea about a year) sayin' t' me, "You're never goin' t' wear fishermen's boots!" But they used t' be smart, yuh know. Yes, they were lovely boots. We used t' wear the duffel trousers as well. They weren't the proper bell-bottoms like you see now (they call 'em flared, don't they?), but they were opened out a bit at the bottom. About 22 inches, I think they used t' be. They never had no flies on 'em neither, just a flap. When they done away with them flaps a lot o' the blokes reckoned they couldn't git used t' the flies! Another thing was a wrapper round the neck, though I never did wear one. That's a funny thing, yuh know, but all my life I've wore a collar an' tie. I can remember when I was courtin' my wife an' she was workin' for Woodcock, the station master, in Denmark Road. I went an' called for her one night an' her boss come t' the door. "She's courtin' a fisherman," he say. "You aren't a fisherman." He said that 'cause I hent got a wrapper on, but I've always wore a collar an' tie. I don't know why. That's just a habit I've got into. A lot o' the fishermen used t' carry herrin' hoom in their wrappers an' they'd take a fair number that way, I can tell yuh.'

The matter of dress was important to the herring fisherman, for it was not only his identifying feature when ashore, but it helped also to compensate for the working hours on board ship when he was clad in his working trousers, his calico jumper, his crotch boots and his oily frock. The walking-out dress of the fisherman, if you like to think of it in that way, was the object of considerable pride as Jack Sturman describes:

'At one time o' day you wore high-heeled boots with fancy toe-caps. They used t' be 19/- t' buy ready made or 21 bob if you had 'em made for yuh. Farman used t' make 'em. He had a boot shop along Waveney Road, where the fish comp'ny offices are now. Then you had the bell-bottom trousers. They were 19 inches at the knee an' 21 or 22 around the bottom. I remember one time when we went t' watch Norwich playin' football somewhere in the Midlands. That was a cup-tie, some time after Christmas, an' they runned a special train from Norwich so a lot o' the Low'stoft boys went. Well, I can remember the girls up there (you know, where we went to) sayin', "Oh, look at their pants." They'd never sin 'em afore, yuh see.

'The coats what you wore had velvet collars on 'em an' they were made o' lovely clorth. That was blue an' heavy. I was once in a Saturday night-Sunday night ship an' when we were laid there at Scarborough about three or four of us went for a drive. They had these charabancs, yuh see, horse-drawn charabancs, where you used t' sit on top in the open. We went out for a country drive, an' we were just goin' out o' the town when we noticed two fellers sittin' opposite havin' a helluva argument. After a little while the one who was nearest me leaned across. "Excuse me," he said, "but will it be rude if I ask how much money you paid for that suit?"

"No," I said.

"Well," he said, "that's what my pal an' me are arguin' about — your suit. He said he didn't think you could afford somethin' like that, so I said I'd ask you."

"Oh yes," I said. "Well, I paid £4-10s an' that was made t' measure."

"Well," he said, "whoever made it didn't rob you of a single penny."

I said to him, "I had it made at Long's, the fishermen's outfitters in Low'stoft."

"That's very interestin'," he said, "because we've made thousands o' yards o' that clorth. An' that was why we were havin' an argument."

I'll never forgit that. That was up in Scarborough. Course, that was a Melton suit, yuh see, an' that really was a lovely thing. You used t' wear a wrapper round yuh neck with it and a cap with a peak down the side.'

The Melton cloth being talked of became a byword in East Anglia for excellence — not least among the farmworkers, who admired its durability and capacity to fence out the rain.

Ernie Armes has a few more superstitions to relate:

'They were a superstitious lot, the ol' fishermen. They'd never start a voyage on a Friday, yuh know. See, when you went away herrin' drivin' you went on a voyage, and that was unheard of t' start away on a Friday. Some o' the ol' boys wouldn't start a voyage if they passed a nun on the way down t' the boat. I know one skipper who's alive today, not far from here, who once done that. He was goin' t' sea one Saturday mornin' an' comin' down the Middle Drive there, when he spotted two nuns. He went back hoom, left his bag an' then walked down t' the market. When he got there he said t' the crew, "We aren't goin' today. We'll go on Monday. A couple o' bloody ol' split-arses just passed me; bloody o' nuns!" Course, the crew didn't mind. They got another weekend at hoom.

'Another thing was that you mustn't mention pigs on board. And you mustn't whistle either. No, you'd whistle the wind up by doin' that! And if a rat come ashore orf a boat an' they saw it — "Bloody hell!" they'd say. "I don't think I'll go on her. She's goin' t' sink." They didn't like women aboard either, an' yit when a new boat went on her trial trip they always used t' have women aboard. They were very blasphemous too, some o' the ol' boys. I had an uncle called Bob Leeds. Swearin' Bob Leeds he was known as an', cor, he was a rum un. He used t' challenge the Almighty, he did. "Come down here, you ol' bugger," he used t' say. "Come down here an' I'll knock your bloody hid orf. If you can't give me more herrin' than this, take 'em back, you ol' so-and-so!" And he'd hull the herrins over the side. Oh, he was a wicked ol' man, he was, an' yit he died aged 93. He never had a day's illness until the last day or two afore he died. Some o' the ol' skippers used t' hull a handful o' coppers over the side. "I'll buy 'em orf yuh, you ol' bugger!" they used t' say. Oh, they were proper hard cases. They bloomin' well were.

'A lot o' the fishermen used t' wear bowler hats at one time o' day. And my ol' man had a blue serge suit and elastic-sided boots with a high heel on 'em. I believe they had fancy toe-caps as well, a scroll on 'em or an anchor or something. He wore a jersey under his jacket an' he always had a silk wrapper go round his neck. I never knew him t' have a jacket with a velvet collar on, but my Uncle Dick an' my Uncle Charlie always had velvet collars. My ol' man's jackets were always blue Melton cloth an' he looked smart in them. He had two gold rings, one in each ear, because that was supposed t' be good for the eyesight. Oh, there were plenty o' superstitions in them days, yuh know. Blimey, if a man went t' sea in them days an' he happened t' have a baby's gaul, he'd never git drowned accordin' t' that. The women wouldn't do their washin' on the day their husbands went orf t' sea. Oh, there were all sorts o' funny ideas.'

The local dislike of seeing those in holy orders caused trouble as late as the 1950s, when the Bishop of Aberdeen, who came down to see the Scottish fleet in action, thought that he would be as welcome aboard the Lowestoft boats as he was on those of his own folk. He found a less than friendly welcome. George Stock can remember similar occasions:

'Some of 'em didn't like parsons. "I aren't havin' one o' them ol' sky pilots aboard," they used t' say. No, the Low'stoft men didn't like parsons very much — except ol' Tupper-Carey. But he was a good ol' boy, see; he was almost like one o' the blokes.

'Some o' the superstitions were funny things. You remember the Bible, where it says that the Lord made all the swine go over the cliff inta the sea an' drown? Well, they reckoned that was why pigs were never mentioned on board ship. I saw it happen so many times when I first went t' sea along o' my brother — this hatred o' pigs. We were out once on the *Rosalind (LT 977)* an' he say t' me, "Boy, did yuh see Cissie?" That was our sister. So I said, "Yes." Then he said, "Did yuh see Ted?" That was her husband, yuh see. I said, "Yes, I went up an' had a look at his pigs." The next thing we knew was bang, bang, bang! — the warps had parted. He never forgot that, my brother, even though that was a wreck what parted us. O' course, we were trawlin', not driftin', but that shows how the superstition was believed. A lot o' the men would bring up an incident like that again and again. We went up t' London once, the wife an' me, t' see a cousin o' mine an' we were all goin' t' Picadilly Circus or somewhere. Hilda an' me were sittin' in the back o' the car, when my cousin's missus turned round (she'd heard about this superstition, yuh see) an' said, "George. Pig,

pig, pig, pig, pig!" And do you know, a bloomin' car come inta us just as she was sayin' it! Thass funny how things happen, isn't it?

'There were other superstitions too. Some o' the ol' boys used t' say that rabbits were unlucky, but I never did pay no regard t' that. I was with a skipper once what dint like the colour green. That was Arthur Cobb, him what hung hisself. We were in the *Ocean Lover (YH 107)* an' he come aboard an' saw that a couple o' the crew had got green tammies. I think they'd bin on a voyage over t' Ireland or somewhere. Well, directly he saw 'em, he hopped down out o' that wheelhouse, pulled the tammies orf their hids an' threw 'em over the side. O' course, some o' the blokes said, "Have he gone mad?" But that was the colour green he dint like, yuh see. All the different men had their different superstitions. What you see with yuh own eyes, you believe in that, don't yuh?

'That was just the same with the gear you wore. There were particular ways with that, I can tell yuh. A pair o' the long boots you wore on board used t' be three weeks' wages. You used t' have 'em made at Farman's or the Yarmouth Stores. I always liked Farman's best myself. After you came hoom from sea you used t' fill them long leather boots up with fresh water an' let 'em stand for a day or two. That was t' git the salt out. Then you used t' put Neatsfoot Oil on 'em. You used t' study your gear in them days. That was just the same with the oily frocks. After the fishin' was over you'd put a broom handle through the sleeves, tie 'em on the linen line an' give 'em a couple o' coats o' Bon Accord linseed oil. You had t' pay for everything in them days, not like now. You never had no rubber gloves or anything then either. My mother used t' make us wool mittens, but there weren't nothin' else. We always had our tan jumpers o' course. Oh yes. The Yarmouth men always wore blue ones. That was strange, that was. "There's an ol' Yarco," you used t' say. Course, I never noticed till I went down t' Shields an' Scarborough an' Grimsby how really in opposition the Yarmouth men an' the Low'stoft men were. I remember readin' in one old book where they used t' almost have fights, the Yarmouth men an' Low'soft men.

'Another thing in the old days was how they used t' bring the Bethel service down onta the market of a Sunday morning. Oh yes, that used t' sound nice down there on the herrin' market. That used t' be just where you went up the first big run-up. And if that rained you'd be all right because you were under cover. You know, that was roofed over. They'd carry the harmonium down, two of 'em, one on each side. Oh, the services used t' be nice down there 'cause all the girls were about. That used t' be a regular thing t' go down t' the Bethel service an' then take a walk round the market. Livin' up this way, in Pakefield, we used t' go to St. George's Hall as well, in St. George's Road. I've got a photo somewhere o' when I used t' play for the St. George's football team. Cor, when you come in from sea an' took orf yuh long boots, you could very near jump over the moon! O' course, you had some good fishermen playin' football for Low'stoft at one time o' day: Joe Thompson an' all them. They were good fishermen an' good footballers. That they were.'

The Bethel, with its robust nonconformity and undenominational status, was always a positive force in the spiritual life of Lowestoft (it still is) and was supported loyally by many of the fishermen and their families. It was originally sited on Commercial Road, but in 1899 the new building on Battery Green Road conveniently opposite the fish market, opened for worship. But even for the faithful the old superstitions lurked never very far away as Ned Mullender recalls in this account of Sunday observance during his boyhood in

117

THE DRIFTERMEN

Pakefield:

'A lot o' the people used t' go to Pakefield Church, especially the wives an' daughters o' the fishermen. The men didn't so much because nine times out o' ten they were at sea. Now those of us what lived the other end o' the village, we used t' go to the mission hall in St. George's Road. That used t' be very well attended an' thass where the Low'stoft town band started from. Yeah, they had a band there afore the First World War an' that was a pretty big band as well. I tried t' join it myself at one time, but with goin' t' sea I couldn't git my band practice in so that weren't no use. Everyone used t' mix in there at the mission. We didn't have no minister; anybody could git up. I remember goin' there one time an' there was an old skipper called Nick Fisher who got up. He was a gret ol' man with a beard an' I can remember him sayin', "Here we are together, brothers an' sisters, arter a long, dreary, ol' West'ard woyage." "woyage," he said — not voyage! Yes, I can remember that. An' there used t' be an ol' gal by the name o' Ruthen. I didn't know her husband or anything like that, but she used t' keep us boys in control o' night time. There was a Sunday school for the young uns there too. I went t' that myself, and when I got bigger an' left school I still went there because all my boy pals an' girl friends went. They used t' vary the services. One time there'd be Brother So-and-so git up an' give a few words. Another time someone else. Like that. My uncle Jimmy, he used t' git up there an' say a few words. You used t' git all the ol' Sankey an' Moody hymns as well. When ol' Nick Fisher finished, he used t' say, "We'll now have that famous ol' hymn, 'When I surwey the wondrous crorss'." Yes, I can remember that right well. You'd have "Lead Kindly Light" an' all them sort. "Keep the lower light burning", that was in the book as well.

'Of course, religion in them days was a funny thing. I mean, my father was a very superstitious man on meetin' a sky-pilot, and so was I. You sometimes used t' see 'em on the South Pier in the summer (you know, as you were goin' out) an' you'd say, "Oh dear, oh dear, oh dear." Another thing was that you mustn't mention rabbits, or pigs, or elephants — an' God knows what else! There were some chaps, skippers, who used t' be against meetin' certain ol' women. Oh yeah, there were a lot o' different items an' dittos. Yeah. Like washin' down. A lot o' the skippers didn't like havin' too clean a ship. I was along of a skipper what didn't like havin' his boat too clean. When there was a gale o' wind, an' there was plenty o' water shootin' about, he used t' say, "Now's the time t' give her a scrub down!" I don't know why it was, but he always used t' say, "A dirty ship is a good ship. That show you're earnin' some money." Well, I suppose that was right if that was scales. I don't mean the cabin an' I don't mean the wheelhouse; I'm talkin' about the deck an' the hold. If you were too particular scrubbin' the hold out, he'd git on to yuh about that.

'Another thing with some o' the old skippers was whistling. No, you mustn't whistle the wind up. Some of 'em didn't even like yuh singin'. Yeah, we had one feller along o' us when I was a young boy an' he always used t' sing on watch. He used t' sing, "I used to sigh for the silvery moon." Ha, ha, ha. My poor ol' father used t' say, "God, no herrin' tonight! He's a-singin' away. Just listen to him." When they shot the nets, they used t' say, "In the name o' the Lord." You'd often hear that said. I never met anybody what used t' buy 'em, but I have heard tell that some of 'em did. They were some what used t' git a bit blasphemous if they weren't gittin' any herrin'. Oh definitely. Some of 'em used t' curse. They all had their own ways an' superstitions an' one thing an' another. Like I said, that was rabbits, pigs, jackoes

CUSTOM AND BELIEF

(monkeys) an' elephants with some of 'em. Then there'd be the ones what didn't like certain colours. Green always used t' worry 'em for some reason.

'Another thing was the srad (shad). You used t' git that in with the herrin'. You'd git an ol' srad or two an' they were quite common. Yes, you got several o' them in the period o' yuh lifetime. Some o' the ol' boys reckoned when there was any o' them about that you din't git many herrin'. I've sin several o' them in my lifetime. They're bigger than a herrin', yuh know, with rougher scales. The colour is about the same, but they're got a different sort o' head. Thass like a bream's head. Where the herrin's is pointed, the srad's is more rounded, if you take my meanin'. But you talk about superstitions, everybody was superstitious at that time o' day. I mean, my wife would git out here in the back garden when there was a new moon. She'd bow to it and chant somethin'. I can't remember it exactly, but the first bit was, "New moon, new moon." Yes, she used t' chant out there in the garden. You mustn't see the new moon through glass, yuh see. I used t' say to her, "Well, you wanta take your glasses orf then." I used t' pull her leg about it.

'Of course, the moon used to affect the herrin'. You used t' git the stronger tides on the full o' the moon, yuh see. The one they used t' call the harvest moon was supposed t' be one o' the best moons for the herrin' industry. You used t' git more fish at the full moon — definitely. Some years a lot o' the boats used t' be down orf the Humber in the early part o' October. Well, if you had a full moon in the latter part o' October, you'd git a lot o' herrin' up south o' the Knoll. You used t' work up orf here east by south, east-south-east, like that. Well, in the November moon you'd out here about 20 miles east-north-east from Smith's Knoll. I'm goin' back a bit now o' course. I mean, the latter part o' the time the herrin' got spotty. You know, there weren't so many about. Well, you know that by how the fleet diminished. A lovely moonlight night, though, you couldn't beat that for herrin'. No, you couldn't beat it. You know, just a little ripple on the water; a bit of a breeze t' make a bit of a stir.'

"Scutchers"

Corton Parish Church at the beginning of the 20th century decorated for the popular 'Harvest of the Seas' service held each year at the end of the home fishing. The pennants stretched above the chancel are the dress flags of the steam drifters *Inter Nos, LT 987* and *Sea Flower, LT 522*, both owned by James and Walter Pye who lived in the village.

CHAPTER NINE

The Drifterman's Year

*'Who'll buy herrings
Fresh and sound?
Who'll buy herrings
By the pound?*

*By the pound
Or by the ton,
Fine fresh herrings
Every one.'*
             (18th century lyric)

There was once a widely held theory that the herring shoals circumnavigated the British Isles in a clockwise direction, thereby giving rise to the different fisheries at particular times of the year. It was an attractive notion, but scientific investigation this century has shown that it was not the case. The most generally accepted idea now (for there is by no means complete agreement as to the species' behaviour) is that there are half-a-dozen or so individual stocks of herring in the waters around our coasts and it was the spring or autumn spawning pattern of these separate stocks that gave rise to shoal activity in different areas at different times of the year. It is easy to see how this gave rise to the idea that the whole herring population was migrating round the coast.

The modern theory better bears out the statement made so often by driftermen that the fish caught on the voyages away from home were "a different sort of herring". Up to a point, this is just what they were. They even varied in size quite often, which meant that the crews had to employ nets of differing mesh size, depending on what area they were working. For instance, the herrings taken in the North Shields season required a small mesh, whereas those round on the west coast of Scotland, at places like Loch Fyne and Castlebay, needed a larger size. Sometimes the boats (the larger ones, anyway) would take two or three different fleets away with them, using what was required at the time and putting the others in store till needed. Other firms would change their vessels' nets by using rail and (later on) road transport.

If one looks carefully at the different herring seasons round the British coasts, it was theoretically possible for the drifters to keep fishing all the year as the following simplified synopsis shows:

New Year — Plymouth or Ayr
Winter to early spring — Dunmore East or Buncrana
Spring — Anstruther or Lowestoft and Yarmouth
Summer — the Shetlands
Late summer to early autumn — North Shields and Whitby, Isle of Man, Scottish lochs and islands
Autumn to early winter — East Anglian home fishing.

The majority of drifters however did not fish all the year round, taking part in all those different seasons, because it was not economic for them to do so. The ones which did were usually the boats commanded by the top skippers, the dons as they were known, men like Jumbo Fiske:

'When the hoom fishin' was over we used t' make up a week or so afore Christmas. Then we used t' go acrorss t' Dunmore. Yes, we used t' go t' Dunmore then. Now I'm talkin' about since the war, yuh know, since the war. Afore the war we used t' go t' Plymouth, but then the herrin' seem as if they all took orf. There used t' be several boats round at Plymouth. Cor blimey, you used t' have yuh Christmas dinner cooked ashore an' then you used t' go t' sea arterwards. That was all right, that was. But I don't know if them Frenchmen killed it, trawlin'. That seem as if they messed it all up. There's never bin any bloody herrin' at Plymouth since they come round there. They were lovely big herrin' as well, just the right ones for the London buyers. That'd be about a seven week voyage alt'gether. Well, then like I say, the Dunmore voyages come along when that finished. You'd go over t' Ireland directly the hoom fishin' was over. Yes, you'd go over there afore Christmas. Sometimes you used t' come hoom for Christmas, yuh know. Yes, come hoom by coach. But o' course you hetta go back agin as soon as it was over. When you were round on that Dunmore voyage you used t' git some very bad weather goin' acrorss near Milford Haven. You used t' land at Milford Haven quite a lot. The Smalls was a bad place round there; The Smalls Lighthouse, about 20 mile out from Milford. Oh, that was a bad place. A lot o' tide an' that, you know. There was one boat lorst with all hands round there runnin' inta Milford. That was the *Shorebreeze (LT 1149)*. One o' the crew even climbed up on top o' the rocks. They found him there, dead, clingin' t' the rocks.

'When I was in the *Sarah Hide (LT 1157)* I used t' go orf down t' Aberdeen in about March. We used t' be right down on the Viking Bank. That was where the best fishin' used t' be, about 200 mile from Buchan Ness. You'd git good trips there an' then you'd follow the herrin' as they moved. They used t' take easterly, half a point o' the compass every week. Well, then you'd git so you were about 220 mile from Buchan Ness. You'd git up an' work right on the edge o' the Norwegian deep water. You'd work up so you were workin' about 220 mile east-north-east of Aberdeen. You'd keep workin' on those grounds for a time. When you were runnin' in an' out o' port you allus looked t' see if you could see anything like. You'd say, "Well, if we can git away tomorrer in decent time, we'll plonk here." You know, t' see if you got anything. But if you dint git anything, well, another time you'd praps drop on to a little shot, yuh see, an' git a night or two's work there.

'When we were up there orf Aberdeen, if the herrin' wun't makin' much of a price, we used t' go an' land 'em in Altona. We used t' shape acrorss t' Heligoland an' then pick the Number One Elbe up. You had t' pick a compulsory pilot up there. You used t' have three pilots altogether. You used t' have a Number One Elbe, another one at the entrance o' the Kiel Canal, an' then the harbour pilot. We did that afore the war, an' we were there in 1939, the year war broke out. An' I said t' the ol' blokes, "Do yuh think there'll be a war?" They say, "No, there'll be no war. We want plenty o' trade with England." That was the pilots, that was. But when we were there at Altona, coalin', the ol' coal bloke said to us, "Not good, Tommy. Too much boom. Too much zoom." He knew all right. O' course, they used t' buy a lot o' herrin' afore the war at Low'stoft, the Germans did. German trawlers used t' lay in the Yacht Basin t' take 'em away. That used t' be solid with horses an' lorries from the market right up Commercial Road. Klondyke herrin', that was. Oh yes, a lot o' German trawlers used t' come over. They used t' cran 'em right into 'em; they used t' salt an' ice the fish down the holds.

'When I went back t' Altona after the war, we got inta port about 12 o'clock one day. I was lookin' out o' the wheelhouse, when along come a feller who was singin' out, "Jumbo Fiske! Jumbo Fiske!" The crew looked an' that was the bloke who was sellin' agent for us when we went there afore the war. This'd be about 1946 or 47 that I'm talkin' about. Well, I dint know exactly how they were sold, the herrin'. We got a price, thass all I know, an allocated price. I spose that was the rules or somethin'. Anyway, they sold them herrins on the market this partic'lar day an' they wanted me t' git up on a box beside this bloke what'd sold 'em. Well, I dint know what they were a-doin' an' sayin', but I did hear 'em sayin', "Sarah, Sarah." I could hear 'em sayin', "Sarah." So I said t' this feller, the agent, "They can speak English. What are they sayin'?" He replied, "They've bin sayin' 'This is the captain o' the *Sarah Hide* who used t' come here afore the war. Hooray! We want t' see more o' him." Well, I couldn't believe it. They were all clappin' an' cheerin'. You're never sin anything like it.

'Afore the war we used t' go through the Caledonian Canal an' come out at Fort William. That was the end o' February, beginnin' o' March, time. Yes, we used t' go right through there an' fish orf Donegal Bay, Teelin, Rathmullen, Buncrana. You'd be close in round there, but that dint make much difference. The Irish never thought much o' herrin' catchin' round there. Well, that was a nice little voyage, that was. Mr. Breach had it all organised properly, yuh know. The senior skipper used t' be in charge. Poor ol' Ephraim Snowling was alive then an' he was in charge o' what you'd gotta do, yuh see. You all had t' put yuh herrin' aboard one boat, yuh see — if there weren't too many, that was. Yes, you'd load one boat up an' send 'em to Ayr. The rest of yuh would stop at sea an' fish. If there was too many herrin', the boat whose turn it was t' go next, they'd put 'em aboard that as well, so there'd be two boats go some days. Then they used t' bring the mail back when they come. They'd bring all the letters an' that, you know. You all went in yuh turn. They'd load yuh up an' you'd gotta go. If the boat comin' back knew you was the next boat t' go, they wouldn't bring your letters 'cause you'd git 'em when you got inta Ayr.

'There was a good few boats in Jack Breach's fleet. There used t' be the *Olivae (LT 1297)*, the *Score Head (LT 120), Feasible (LT 122), Girl Gladys (LT 1174), Strive (LT 133),* the *Swiftwing (LT 675), Three Kings (LT 517)*. Oh, there was a proper fleet of 'em, yuh know. That was good organisation round there too. When you were runnin' the herrin' inta Ayr, they used t' give yuh one box for every two cran you had on board. Say one o' the boats had caught 20 cran, well she'd give you 10 boxes o' fish for carryin' 'em. Understand? That was what you'd git for takin' 'em in. That was a sort o' commission an' that used t' go inta yuh earnins. You'd git that from all the boats an' that was t' make up for you not catchin' any herrin' time you were runnin' everyone else's inta Ayr. Like I said, you had one skipper in charge (Freezer Ellis was another one what did it) an' he used t' git on the radio an' let 'em know you were comin' an' how many cran you had on board. There used t' be some rum ol' passages too round there; some real dirty ol' passages goin' acrorss. You know, round the Mull o' Kintyre. Oh, thass a nasty ol' place! — a lot o' tide an' bad weather. They sold there Sundays an' all in Ayr. Yis, they sold there Sundays. Thass the honest truth, that is. Nelson Utting was our salesman round there. He worked for Hobson's.

THE DRIFTERMEN

Oh yes, they landed Sundays all right.

'We were on that voyage several weeks too. That we were. We'd start orf about the end o' February time an' when we got round there that'd be about the first week in March. We used t' fish round on that west side, yuh see, till they opened the Shetlands in the latter part o' May. The best herrin' we used t' catch was round orf Castlebay an' up there. Cor, beautiful herrin', they were! Beautiful herrin'. You used t' lose a lot out o' yuh nets if you dint have a very big mesh, yuh know — unless you got yuh baler t' work. Oh, they were beautiful herrin'. We've didalled 'em as well, many a time. And used a thief net. When I was in the *Sarah Hide (LT 1157)* we used t' run one o' them up. You'd have a pole go out orf the top o' the wheelhouse an' the net used t' hang down orf that alongside the boat. That had a trapstick on the top part o' the net so the net dint close up an' then the bottom used t' go down inta a kind o' cod end. Coo, we used t' git a lot o' herrin' out o' that. They used t' fall in as you were scuddin'. Yes, they were good things they were, them thief nets. They made a big lot o' difference to yuh, they did. They'd fill some boxes up, they would.

'Once the Shetland fishin' started you'd go there an' have a go on the different grounds. You'd try the Brassy Shoal (Bressa Shoal) an' work all round there. We even used t' run the fish inta Aberdeen sometimes if that suited us. Then you'd be about 15 or 16 mile out from Bard another time, praps 20 mile. Sometimes you'd be up orf Sumburgh Head. Since the war the herrin' always seem to have bin a lot closer to the land down on the Shetland grounds. Afore the war you allus used t' go 50 or 60 mile, or 40 or 50 mile, out from Bard. East-south-east you'd be. Bard is a big corner o' land stickin' out. All the boats were workin' out o' Lerwick. Sometimes, when we dint have much herrin', we'd go inta Lerwick an' land, git a clear ship, store up with coal an' that, then go out an' git a trip so we landed in Aberdeen on the Monday. That was a good market in Aberdeen on the Monday, but that wun't no good at Lerwick. The Scotchmen dint fish o' Sundays, yuh know. No, they didn't. When you landed at Lerwick, you used t' run the herrin' up on the ol' bogies. They were a kind o' trolley an' you used t' put about two cran a time on. That was hard work, but when you were young an' used t' work there afore the war you dint pay no regard to it. You were young! You used t' have a laugh an' a yarn with the Scotch girls, yuh know. A lot o' them guttin' girls used t' be livin' up the huts there in Lerwick. They were famous, them huts. All sorts o' goins on there. But them girls used t' work hard, yuh know. Coo, they did work hard! They were very quick with their hands.

'After some o' the grounds slacked orf we used t' come up orf Sumburgh Head an' Fair Isle an' work up that way. Nearly all the boats used t' work from Lerwick all the while; they'd be there till August. But when I was in the ol' *Sarah Hide* I used t' git back up orf Aberdeen agin. You know, when Shetland began t' git a bit thin. I liked that Aberdeen voyage. That was a good voyage, that was. When we were gittin' near the end o' the voyage, we used t' draw up t' Shields praps for a week. Then we'd generally go inta Scarborough Wake on the last night afore we went hoom. You'd shove the little boat out. You know, go ashore in the rowin' boat an' git some rock. Yes, all the boys used t' git some rock t' tairke hoom for the kids. That'd be about the second week in August, or the third praps, an' you allus reckoned on bein' hoom well in time for Low'stoft Regatta.'

What a splendidly matter of fact account of one man's Odyssey! Although as we have seen all the driftermen did not circumnavigate the British Isles annually in their search for herring a great many made the Scottish voyage and these have never been written about as much as the Scottish men's annual invasion of East Anglia.

Horace Thrower recalls what these voyages were like:

'You used t' git some good hauls orf Lerwick in the Shetland Isles. Oh yes, you'd git a good lot there sometimes. You used t' notice up there that one time you'd go outside an' git herrin' in deep water; the next time you'd catch 'em where that was shallow. Say you went 30 or 40 mile out — well, you'd git herrin' that night an' you'd come in. The next time you went, praps you couldn't git that distance so you'd go an' shoot about 10 or 12 mile from Sumburgh Head or anywhere like that. Well, 9 times out o' 10 you'd git herrin'. One night in deep water, one night in shallow. I noticed that time arter time. See, you'd go out an' chance yuh luck. If you couldn't git away, that'd be dark afore you got very far so you used t' shoot close in. See, you're gotta have yuh nets in the water afore dark as a rule. You shot with the close if you could. That's when the herrin' rise, just after the close. When you come in with a shot up there you just landed 'em on the stages at Lerwick. Yeah, the stages. You used t' have a bogie drawin' 'em up t' where the girls were guttin'. You'd git about two cran on at a time, about eight baskets, an' you'd run 'em up. Course, that all depended what sort o' stage yuh got. Sometimes you got a good stage where that was pretty level. But then you'd git another one where you hetta go uphill with 'em, push 'em uphill all the time.

'That was about a 10 week voyage up there, I spose; 10 or 12. That all depended, yuh see. If the herrin' lasted, you'd stop there. But if the herrin' took orf, then you'd come down t' Fraserburgh an' Aberdeen. The Moray Firth, thass where you git the jellies. Coo, thass about the worst place there is on the east coast for jellies. See, you shoot at the mouth o' the firth an' I spose they come out from the firth. Thass the only thing I can think. Oh, I dint like them one little bit! By the time you got down t' Shields you got a smaller herrin', and o' course there was a lot o' boats what used t' go t' Shields only. Work out o' there all the time, see. They had a smaller mesh net there, where down at Shetland you got a bigger herrin' an' so you'd have a bigger mesh net. Most times when you went down t' Shetland you'd take a new fleet with yuh. You'd take a new fleet an' you'd tan 'em at the weekend every so orften. Yeah, you'd shoot for about a fortnight an' then you'd tan 'em. You'd do that on board when you were in the harbour.

'I have heard tell o' some boats tannin' nets as they went along, but I've never done that myself. That was a bit too risky, tannin' nets as you were steamin' along. No, I dun't believe in that! But you know what some o' the skippers were like — they were buggers! No, we allus tanned ours in harbour. See, sometimes down at Shetland, if you'd had a good shot on the Friday night, you'd start tannin' on the Saturday mornin'. Then you'd hafta let 'em lay all night after you'd finished an' git 'em all sorted out on the Sunday. You used t' be down on the perks when you were tannin'. The ol' skipper would stand there in the wheelhouse, a-timin' yuh. You'd give 'em three minutes a net an' you could do about half a dozen at a time. There'd be two men on each side o' the tank a-rammin' 'em down an' then they'd pull 'em out after about 3 minutes. You used t' have a kind o' shute runnin' from the fore-side o' the wheelhouse down inta the tank an' you'd pull the nets up onta that as they come out o' the tank. Then you'd let 'em drain on that for a little while, yuh see. When the skipper told yuh t' take 'em away you used t' stack 'em along the side o' the ship — all along in the kid.

'You used t' have an odd number o' nets on board. About 81, 91, 101, somethin' like that.

Oh yes, you must have that odd one. They used t' stretch a fair way when they were out an' they'd take some haulin' too. Yes, you'd have all hands after 'em then. I remember once when we were out orf Aberdeen. We see the sun set an' we see the sun rise — an' we were still at work! We musta got 300 cran that time. The weather was bad an' we had a hard job t' see the nets. You know, now an' agin as we were haulin' the buffs would pop out o' the water. Just the odd one would; all the rest were under. That was proper blinkin' bad, that was. There was so much swell down there, yuh see. You git deeper water an' you git more swell. I tell yuh, we see the sun rise an' we see the sun set — an' we were still at work. That was in the *Norfolk Yeoman (LT 137)* along o' Doff Muttitt after the war.

'The Scotchmen wouldn't fish on a Sunday, yuh know. No, they were all Saturday an' Sunday night boats. They used t' go out after midnight on Sunday night. When you were down there, in their ports, you couldn't fish Sundays either. No, you couldn't go out on a Sunday; you had t' wait till after midnight. The only thing you could do was go out on Saturday an' stay out. The Scotchmen still kept their Saturdays an' Sundays when they come down here on the hoom fishin', but we dint o' course. We just used t' keep the Saturday night an' go out first thing on Sunday mornin'. There were still several sailin' drifters left up at Lerwick in my time. That was the only place you used t' see any quantity o' sailin' ships. They used t' come down here years ago, but that was afore my time. Yis, I can't remember them comin' after herrin' down here (I dint live in Low'stoft, yuh see), but you used t' see 'em in the Shetlands. Then they gradually faded out. When I last went down there, I don't think there was any, but when I first went there was. They used t' be level with the rail. No sides at all, just like a blinkin' barge! I shouldn't like to have bin on them. Gret ol' masts an' sails! O' course, the latter part o' the time they had motors put in 'em, yuh know. They done away wi' that big sail an' had motors put in. They kept the mizzen, but they got rid o' the big sail.

'We used t' git round on the west coast as well. Stornoway, Oban, Ullapool, all round that way. That was after the war, that was. They were a good herrin' round there, big, bigger'n what they were on the east side. Yes, you needed a big size mesh round there. Specially at Stornoway; they used t' be massive herrin' there. We used t' go through the Caledonian Canal, or through the Pentland Firth an' round that way. They were the only two ways you could go. I never went on a full voyage round the west side like some o' the boats did (they'd be round there time you were at Lerwick); I only went when there wun't much doin' on the east side. I remember one year goin' t' Wick. That was when I was in the *Patria (LT 178)*. There wun't many herrin' on the east side an' some ol' Wick skipper say t' our skipper, "Would yuh like t' try somewhere else?" Our skipper say, "Yis, I'd like t' have a go." So the other ol' boy, he say, "All right, I'll take yuh through the Pentlands." Well, that was the first time we'd bin through the Pentlands, so we follered him through an' shot orf what they call Whiten Head. There was another Low'stoft boat with us, but I can't remember her name. We shot about seven or eight mile from what they call Whiten Head an' we got a lovely shot o' herrin', over 100 cran. But him just up t' wind'ard of us, this other man what'd shot down at our pole-end, he was full up with dogs. He spoilt a whole blinkin' fleet o' nets. We landed in at Wick an' we went out the next day, went out on the same ground — an' we got dogs! We spoilt a whole fleet o' nets an' we had t' change 'em. They sent a new lot from Low'stoft up t' Wick an' we sent the other lot back. We couldn't do nothin' with 'em. They were too bad.

'I went round t' Dunmore as well on that west side fishin'. The first year we got married, I think it was, 1927, I went round there an' I was gone over 20 weeks. We started at Newlyn for mackerel an' then we changed our nets there for herrin'. We went t' Milford Haven, an' then from Milford Haven we went acrorss t' Dunmore on the Irish side. We were there for several weeks, then we changed our nets agin an' went further north. We went up t' Oban, then from there we went t' Stornoway. We fished there for a bit, then we went through the Caledonian Canal an' come up t' Aberdeen, Fraserburgh an' Peterhead. We were gone over 20 weeks alt'gether. People wouldn't believe that today, but thass the truth. We changed our nets three times that voyage an' when we finished up we never had a penny t' take. We just cleared ourselves. No profit. You wouldn't believe it, would yuh? All that time away, 21 weeks! Oh, that was a blinkin' hard life. You went all that while an' had hardly any money in yuh pocket t' spend. O' course, you used t' git yuh weekly allotment. That was 16 or 17 bob at that time o' day. That used t' be sent to yuh wife by registered letter an' the company used t' deduct the postage! Thass how they used t' treat yuh! I was in the *Empire's Heroes (LT 703)* then. An' do yuh know what? — we come hoom, changed the nets an' went back down t' Shields. Thass right, that is. We had just a few days at hoom t' tidy the boat up a bit an' then we were straight back down onta the North Sea.'

Ned Mullender was another who took the long voyages. His yarn, like the ones above says more about the drifterman's itinerant life than statistics ever could:

'Yuh first voyage years ago on the herrin' was round t' Plymouth just afore Christmas. You used t' go there after the hoom fishin' an' you'd be round there till about February time. Well, that all depended on what fish you got o' course. I'm goin' back a lot o' years now, yuh know — 1919, 20, 21. They were lovely herrin' round there at Plymouth. They used t' make a lot o' money. I know when I was in Joe Colby's boat, the *Silver Herring (LT 1145)*, we got 12 cran round there one night an' they made over £7 a cran! Now thass goin' back to about 1920. You used t' go inta Plymouth an' land 'em right in the little ol' dock there. One o' the buyers was Joe Jary an' they used t' send 'em up t' London mostly. Well, they'd send 'em anywhere they had any orders, but London was the favourite place. Ol' Bill Champion was another buyer round there. He was a chap who weighed about 20 stone an' he'd be there. Well, he used t' foller the boats all round. He'd be at Scarborough too when you went there. There wun't right a big lot o' boats went down t' Plymouth, but there was several. I couldn't tell yuh exactly how many, but you'd git Low'stoft an' Yarmouth crews mainly. You only shot a small fleet o' nets round there at Plymouth. Yes, you only shot about 45 nets, praps 50.

'Well, that was the start o' the herrin' fishin', goin' down there just afore or just after Christmas. And after that voyage finished a lot of yuh would come hoom, take on mackerel nets an' go back down t' Newlyn for the mackerel season. Well, them what dint go mackerel catchin', they'd stay at hoom till the Scotch voyage begun. They'd pay orf, clean up the ship an' lay until they were ready t' go. There used t' be a season springin' out orf here at one time, yuh know. I think one o' my namesakes what used t' live up the road here used t' go. He had a little ol' Yarmouth boat called the *Kernoozer (YH 754)* or somethin' like that. Yes, he had her, an' he was out here springin' one year an' he got a little voyage. That was quite early on. Mighta bin about 1920, or just after the First War. O' course, there

127

used t' be several boats go springin' afore the First War. Out o' here an' out o' Yarmouth. Oh yes, they'd be out after the spring herrin'. They were all spents, o' course, but still there was nothin' else about.

'As regards the Shetland voyage, we'd work down there about a dozen weeks, I spose. Yeah, about 10 or a dozen, an' then we'd work our way up. You had two or three species o' herrin' down there. You'd git a small herrin' for a start an' then you'd git a bigger herrin', quite a nice one that was, an' then the latter part you'd git some spents in. To begin with, you'd probably be workin' out o' the north entrance o' Lerwick Sound. See, Lerwick is on the main side, then there's Brassy Island, an' there's an entrance t' the north and an entrance t' the south. Well, you'd work out o' the north entrance the first part o' the time an' you'd vary a point o' the compass as you worked. Say the first two nights you'd be out there about east-north-east; well, the next night you'd be east by north. Then next time you'd be east from Score Point. See? You'd start in about 24 or 25 mile an' then you'd git to about 50 mile out, praps more'n that. See? Well, then the next week someone would probably go out o' the South Entrance an' go east from Bard. If he got any herrin', all the rest o' the fleet would soon be down there. That they would! There wouldn't be no room anywhere, so you'd try somewhere else. Praps you'd go east by south, only instead o' goin' 20 mile out you'd go 30.

'You used t' git a lot o' dumpin' at Lerwick. My brother-in-law's brother in the ol' *Cen Wulf (LT 49)*, he dumped about four or five hundred cran one week. He couldn't sell 'em. No one dint want 'em. I mean, you had Scotch boats an' English boats all workin' there. Yes, there was a lot o' boats workin' Lerwick one way an' another. I expect you're heard a lot about the huts an' all the goins on there, haven't yuh? Well, a lot o' that is far-fetched. You used t' go in an' have a sing-song, but apart from that there wun't a lot of other stuff. No, no, not up there. A lot o' them girls were real nice girls. When we got down there in the *Reunited (LT 68)*, the first year we went, we laid at Slater's Stage. On Sundays after we'd coaled an' watered, we used t' invite the girls aboard. You know, we used t' coal an' water on the Friday, if we could. Then we'd be all right. We wouldn't hafta lay at the market. See, the boats used t' lay broadside t' the market an' you could be four or five boats out. Yes, there'd be rows of 'em the whole length o' the place. So them what could git t' lay at a stage, they did so. Yes, I looked out t' git coaled an' watered on a Friday if I could, an' then on Saturday we used t' go right t' the stage, land the few herrin' what we had an' lay there all the weekend. Then we used t' invite one or two o' the Scotch gals aboard t' have a Sunday dinner. You know, proper respect an' all that. Then we used t' meet 'em in Low'stoft when they come up here. They were lovely girls, an' they even used t' come along on Saturday an' lend yuh a hand all right. They'd fill the needles an' mend. I used t' take me ol' mandolin along on that voyage an' we used t' sit round an' have a sing-song.

'After you'd finished at Lerwick, you'd come acrorss t' the main. You'd probably go inta Aberdeen the first weekend — either there or else inta Peterhead or Fraserburgh. That all depended. You wouldn't work out o' Aberdeen for very long. You'd work yuh way down from there, down along the east coast. You'd work yuh way up t' the Longstone, shoot at the Longstone, then come up t' Shields. You'd still have yuh Scotch fleet o' nets on board. You know, what you went down with. You dint always bother t' change 'em, but o' course you wouldn't catch so many herrin' as what the Shields boats would. Some o' the Low'stoft boats never went no further than Shields. They had a small mash net an' they used t' work out o' Shields the whole voyage. Where a net what you went t' the Shetlands with would be about 30 t' 31 mashes t' the yard, the Shields would probably be about 36 or 37. See? So you got a lot o' herrin' go through your nets. The only thing you benefitted by was that you got a little better price t' make up for it, because your herrin' were bigger. Once there, on the way down, we shot orf Shields an' we got about 18 cran. Now the boat alongside us got about 40. That was the ol' *Briton (LT 1017)*, that was. When we got in, he got about 17 bob a cran, where we made 21.

'You used t' go inta Scarborough as well, o' course. When I come out o' the First World War I was along o' George Rushmore in the *Hope (LT 1075)* an' we used t' go an' lay at Scarborough nearly every weekend we was there. That was very rare we'd miss. We were workin' out o' Hartlepool the rest o' the time. He liked workin' out o' there because that was partly his own boat an' his own gear an' you dint git knocked about so much as you did at Shields. You know, at Shields there was rammin' an' pushin', but at Hartlepool there wun't so many ships. And you'd got a nice dry dock t' go in too, so we used t' work out o' there. Well, come the weekend, he'd say, "I think we'll have a night at Scarborough." So that was what we used t' do, go an' have a Saturday night there. O' course, comin' a bit further down, there used t' be a big school o' herrin' at the Dowsing. Thass a spawnin' ground there. The herrin' are all maisy. Oh, my word, yes. And I know once (I'm goin' back now afore the First World War) we got a couple o' very nice shots there one week. They were both over 200 cran. And that was out of about only 65 nets, that was! You dint used t' shoot much more'n about 65 then. Well, o' course, when they started agin after the war they began t' lengthen the fleet. They went t' 81 an' then 91. Some of 'em woulda shot 191 if they could, I reckon! You used t' go inta Grimsby orf the Dowsin'. You'd mess about there an' orf Flamborough Head. Then, when it was time t' come hoom you'd come. You'd have about a week at hoom, cleanin' up an' squarin' orf an' gittin' a fresh lot o' nets, then you'd go away agin down on the North Sea till the hoom fishin' began.'

And the last word about the life of the wandering driftermen comes from Herbert Doy:

'Some o' the boats used t' go round t' Plymouth afore Christmas. I've bin round arter herrin' myself, shootin' orf the Ritz. You can lose some nets there too if you dun't watch it. Dutch Turrell used t' be the boy for fishin' round there, that he did. Sometimes you could go out round there, shoot yuh nets, haul 'em in an' be back in agin in time t' go t' the pictures. You'd shoot in the afternoon just afore tea (that got dark early, see), git a little shot o' herrin' an' be back in agin in time t' go t' the pictures. Then you'd gotta set a watch on the herrin' what you'd got aboard. You know, till next mornin'. That was because the sharks would be aboard a-pinchin' 'em. The Plymouth blokes, I mean. Yeah, you hetta set a watch in the wheelhouse arter you got in. All the herrin' lay in the kids, yuh see, an' you'd see all these here blokes come crawlin' along, boys an' all! They were after pinchin' a bag o' herrin'. Yeah. O' course, you used long strops on the nets down there 'cause you were in deep water.

'Durin' the First War I had a voyage round at Stornoway. That was in 1917, when I was along o' Ted Chilvers. All the boats fishin' then were the ol' crap, the ol' muck. People wun't go away in 'em nowadays. We were in the *Gannet*. She had a big ol' iron wheel. I can remember that. We went through the Caledonian Canal t' git there an' we hetta have an ice-cutter go up ahid of us. Yeah, we had an ice-cutter go ahid of us t' cut the ice. That was just afore Christmas an' we got t' Stornoway at the time when the town hall caught fire. They never had a proper fire engine; they only had a little ol' wheelbarrow sort o' thing. That burnt right down, the town hall did. Yes, that was the Christmas time o' 1917 when that caught afire. O' course, a lot o' people reckoned that there was a lot o' twistin' goin' on up there, so someone wanted t' burn some o' the papers up, I reckon, afore that all come out. Anyway, we were up there an' there was barrels o' whisky standin' around. Everyone was fillin' bottles up! Yes, we all were. When we went back at the end o' the voyage, we went through the canal agin. "All right, skipper," the pilot say, "let go o' the wheel. She know her way through here." Course, he was only jokin'. "She's bin through here enough times," he say. See, she belonged t' Shields when I was in her. Anyway, we went aground on the 22 mile run through there. Yes, that we did. We went aground on the 22 mile run an' we carried our nets ashore orf the boat. Ol' Whistler Stammers in the *Research (LT 357)* was with us an' he tew us orf. He got us orf an' we'd lorst a fang an' a half orf the propellor. An' thass how we come back t' Aberdeen, like that. We come back with two fangs an' a half on. She was a-rattlin' the whole bloomin' time, I can tell yuh!

'Mind yuh, we got a decent voyage round there. That was the same year the *W.A. Massey (LT 1090)* got lorst. Yeah, the *W.A. Massey*, lorst with all hands. That was our last night fishin' an' we were all goin' t' come hoom together. That was our last night fishin' there at Stornoway, when she got blew up. We heard the noise, but we never did see nothin'. Eventually they picked up some o' her baskets lashed t' half o' the small boat. That was a mine what blew her. We dint see it. They just picked up half the small boat with baskets lashed onta the side. Thass where you used t' lash yuh baskets, yuh see, on the side o' the little boat.

'I have bin round on that west coast two voyages a year; one at Christmas time an' one in the summer. They were lovely big herrin' round there. You only used t' shoot about 28 nets, so you know what sort o' size they were. We once got 111 cran out o' 28 nets. Yes, that we did. We were a-landin' 'em an' that come on Saturday night. Well, the Scotchmen knocked orf at midnight; they dint work Sundays. So we just hetta chuck a little salt on what we had left an' git 'em ashore after midnight on Sunday. One o' the Low'stoft skippers, Charlie Sampson, lorst three fleets o' nets that year with the weight o' fish. Yeah, he lorst three fleets o' nets an' then he started t' double-buff 'em. We used t' go loch fishin' as well on that Stornoway voyage. You'd only shoot a few nets, anchor 'em, an' then go an' shoot a few more somewhere else. You used an anchor and a dan on that job, an' the wash would keep the nets orf the rocks. There was quite a lot o' loch fishin' round there, specially Ullapool way. I've landed herrin' in there more'n once. You'd drop two or three lots o' nets out, praps seven or eight a time, praps a few more. You dint leave 'em for too long an' then you'd go back an' haul. Course, there's a bit o' fresh water attached t' them lochs, but they were bloomin' gret herrin'. The Scotchmen used t' go an' anchor their ship in the lochs for the weekend. They used t' leave the ship anchored an' go hoom! That was all up Ullapool way.

'They were the normal sized Scotch nets you used, only you dint work a rope, a warp. The water wun't all that deep, so you dint want the warp catchin' on a bit o' rock an' pullin' the nets down. You used t' leave 'em loose at the bottom. We learnt that trick orf the

Scotchmen. Well, we had a Gaelic pilot along o' us when I was with Ted Chilvers. He used t' show us what t' do. I was on watch one night an' I see all these rocks an' the wash breakin' agin our stern. I went arter him, this here Gaelic bloke, an' told him. He say, "You're all right. You wun't hurt here. The wash will keep yuh away from the rocks." That was right an' all. See, he knew his water. The Scotchmen always did do that loch fishin', but as for Low'stoft boats I can't say. We did during' the First War. I was round there when them 19 cargo ships all got sunk. The submarines got betwin the escort an' blew the bloomin' lot up! Oh, they got round there all right, the submarines. Good God, yes! We never had no trouble, though. No trouble at all. The winter voyage round there lasted about 8 weeks an' the summer one about 10.

'I was round at Dunmore three year runnin' durin' the Second World War. That was with ol' Jack Wiley in the *Golden Ring (LT 593)*. That was Christmas time an' all. The voyage would last about a couple o' months, I spose, an' you used t' run the herrin' acrorss t' Milford Haven. If you'd only got a little, like four or five cran, you used t' go an' land at Dunmore or take 'em up t' Waterford. One year we were round there, ol' Shipsy (he was a sort o' politician bloke what lived at Waterford) cooked our Christmas dinner for us. Yeah, I was round there in 1940, 41, 42. When we finished at Dunmore we used t' mess around for a bit, then we'd go t' the Isle o' Man. That was the Isle o' Man summer fishin'. That lasted all summer. You'd shoot orf Port Erin, orf Douglas Head. Or you'd run from Peel. We used t' run about six mile from Peel an' then shoot. After Dunmore finished we dint used t' go hoom, yuh know. We used t' send our nets hoom by truck an' git fresh ones back. The Dunmore herrin' were a lovely big herrin', but the Isle o' Man ones were smaller. Mind yuh, that was a good fishin' there. You'd go out late at night, just afore dark, an' shoot; you'd be back agin about 3 o'clock in the mornin'. Some o' them ring-netters round there used t' fill their bloomin' selves up in about an hour! Thass when they all started t' come round there, the early part o' the war.

'We used t' shoot about 73 nets on our boat, an' I know once we got 160 cran there an' brought 'em inta Fleetwood. We landed 'em in at Fleetwood from the Isle o' Man. We were the only boat what caught the gates, the lock gates. We got hauled quicker than all the others an' we come acrorss an' just caught the lock gates afore they closed. We got the control price for our herrin'. That was the full fixed price — £4-18-0d a cran. That was in 1942, in the *Golden Ring*. There was plenty o' dogs round there on that fishin'. Dogs an' jellies. Cor, you used t' git stung t' death! They'd bloomin' near blind yuh. You used t' turn yuh sou' wester round an' shake the nets so the jellies dint come in yuh face. Oh yes! You hetta turn 'em out o' yuh net. You'd run the gill up, the side o' yuh net, an' turn 'em out. We used about 34 meshes t' the row round there. 34 or 35. So that was a mesh just about over an inch square.'

Horse Mackerel or 'Scad'

Above, the crew of *Shipmates, LT 1134* pose for a photograph while round at Newlyn for the mackerel fishing. Such photographs were popular with driftermen just before the First World War. *Shipmates* was a steel steam drifter built by Fellows of Yarmouth in 1911. Below, some of the abundant herring being made up into what where known as 'rough packs', that is shovelled into wooden barrels and salted down.

CHAPTER TEN

Cousin Mackerel

*'The herring loves the merry moonlight
And the mackerel likes the wind.'*
(Traditional)

The driftermen had only one important supplement to the pursuit of herring and that was the mackerel fishing. The mackerel too is a pelagic fish that swims in shoals and can be taken in drift nets. The most important mackerel fishery was the Cornish one, with the bulk of the fleet being based at Newlyn, but there were also important North Sea seasons at one time in the spring and in the autumn. These latter diminished greatly after the First World War but between 1890 and 1910 the Great Eastern Railway ran special Sunday trains to Birmingham and Manchester with catches of Lowestoft mackerel. Most people in Lowestoft and Yarmouth however never rated the mackerel very highly and there were often considerable dumpings of the fish during times of heavy landing. The herring was very much king and the mackerel had a dubious reputation. It was reckoned to be a 'dirty' fish, even poisonous, and this was one reason for its unpopularity among the East Anglians. It was a rather unjustified view as some of our witnesses will show.

Mackerel feed largely upon small pelagic crustacea (especially copepods), except during the winter months when they tend to lie in gulleyways or in deeper water just off the continental shelf. At such time there is little movement in the shoals, and this contrasts strongly with their activity at other times of the year, when they swim in large shoals close to the surface. Spawning takes place during the spring and summer in shallow water and can even last through till the early autumn. Out in the North Sea one sure sign of the presence of mackerel, as far as the fishermen were concerned, was a brownish tinge in the water. It was popularly known as hoss pissy water, but while it augured well for mackerel it was no good at all for herring.

The method of catching mackerel was similar to that for taking herring, but there were differences in the nets and in the way they were worked. For a start, with mackerel being larger fish than herring, the mesh size of the nets was larger, 24 to 26 rows per yard being about the norm. The nets themselves were considerably smaller, having a length of about 30 yards when stretched and 21 when set in and ready to work. Their depth could be anything from nine score meshes to 15 score. The net-rope had large, oblong corks right along its length and these were sufficient as a rule to keep the nets floating up near the surface of the water, which was where they hung. The warp passed underneath the fleet of nets as in herring catching, but unlike the latter method the bottoms of the nets were not secured to it. Instead, they hung loose at this end of things, with the seizing running down directly from net-rope to warp (see diagram).

During the 20th century there were two types of mackerel nets in use, the rough net and the Scotch mackerel net. The former was the old, traditional one, descibed above, made of hemp twine at first and later of heavy cotton. It was customarily treated with linseed oil to give it long life and Flip Garnham recalled his firm's practice of laying out mackerel lints

THE DRIFTERMEN

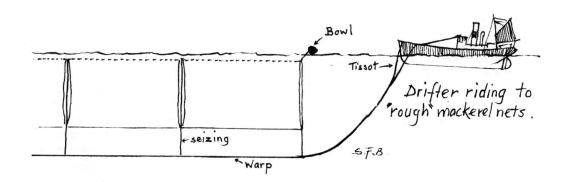

Drifter riding to "rough" mackerel nets.

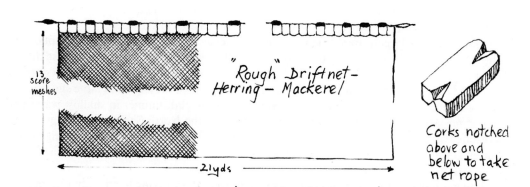

"Rough" Driftnet – Herring – Mackerel

Corks notched above and below to take net rope

Flat corks – close spacing where extra buoyancy is needed – ends, "rough" mackerel nets, etc.

on the autumn stubbles near St. Margaret's Church after they had been done. This was so they would dry off, aerate and not heat up to a temperature where spontaneous combustion became likely. One happy by-product of this spreading out was the number of pheasants and partridges that got ensnared in the meshes, thereby becoming the 'property' of the ransackers when they arrived at the field early in the morning. The Scotch mackerel nets were of finer cotton and generally lighter than their predecessors, and by the 1920s their use was becoming widespread. They got their name because they were like the Scotch herring nets, though not so deep, and it isn't unusual to come across the term mackerel-herring-nets, strange and paradoxical though it may seem (for the rig and working of Scotch mackerel nets, see the diagrams in chapter 2).

The annual spring migration to Newlyn and Penzance goes back to the 1860s, when the East Anglian sailing drifters first began to make the Cornish fishing a regular, annual event. It was a voyage that usually lasted from the end of February through till May and, because there were profits to be had, more and more boats tended to go. Naturally enough, the Cornishmen began to resent this influx of 'foreigners' every year, and not just because they were muscling in on a traditional means of livelihood. The Yarmouth and Lowestoft men also insisted on fishing and landing catches on Sunday, which was a recognised rest day in the south-west. This was not solely due to religious feeling; it was also reckoned to give the mackerel themselves a rest for spawning purposes, thereby maintaining stocks. Ill-feeling between the two camps festered and grew for a number of years, until it ultimately flared up into trouble of a sufficiently serious nature for the government to intervene. This was the Newlyn Riots of May 1896. Fighting took place around the quay and on board the boats and troops were called in from Plymouth to stop the disorder. Arbitration eventually brought an uneasy peace. It was known as the 'Berrington Compromise' and by its terms each side conceded things to the other without being really happy at doing so. But that wasn't the end of the matter, because for some years afterwards it was customary for a naval gunboat to keep an eye on the fishing, just to make sure that it went smoothly. For instance, a photograph in 'The Sphere' magazine of April 20th, 1901, shows *H.M.S. Rattlesnake* lying just outside Newlyn harbour. The caption mentions, among other things, the increasing number of steam drifters then beginning to go on the Westward voyage — boats, it says, that were 60 to 70 feet in length and some 30 tons net. Things were more peaceful by the time Horace Thrower took his first mackerel voyage:

'When you come hoom orf that Plymouth voyage, you used t' go back agin down t' Newlyn, mackerel catchin'. That was what we called a clean voyage. You know, you aren't messed up too much. All the mackerel you git, you pick 'em out one at a time, yuh see. You can't scud 'em out like you can a herrin'. If you got too many, you'd leave 'em in the nets. You'd what you call run 'em in, pile 'em all up an' clean 'em when you got inta port. I mean, you'd take so long time you were at sea; that took a long while t' pick mackerel out. They were rum things t' handle. I've had my hands blown up an' blood runnin' out o' the ends o' the fingers where I've bin pickin' of 'em out. We dint have no gloves then, yuh see. Gloves weren't thought of then like they are now — not rubber gloves. Yuh hurt yuh hands when you clip the gills gittin' the fish out o' the net, orf the mashes. Thass how you do it, see: you clip 'em so they just fall out. When I was in one boat, I know we come in there once an' we had so many mackerel that we'd got nets all round the boat. That was piled up with 'em. The skipper could only just see out o' the wheelhouse 'cause there was so many mackerel. An' how many hours we were cleanin' the nets in harbour, goodness only knows. That was round at Newlyn.

'When you did that fishin' you had what were called Scotch mackerel nets an' also the ordinary rough nets. Some o' the boats used t' shoot a mixed fleet. You know, there'd be praps a quarter fleet o' rough nets at the pole end and then you'd have Scotch nets the rest o' the way. The rough nets were joined together down the sides so they were like one long stream, but the other ones were worked more like a herrin' net. They gradually done away with the rough nets until you had all what were called mackerel-herrin'-nets. They weren't as deep as a proper herrin' net an' they float on top o' the water. They don't sink; they float. They had the big flat corks on 'em, an' then every so often you'd have a bunch o' corks or little tiny buffs t' act as floats. By the time you'd got them out, there was well over 2 mile o' nets out ahid of yuh. You put out a larger fleet for mackerel than you do for herrin'. See, when you're herrin' catchin' you've got yuh perk boards where yuh nets lay on. Well, for mackerel catchin' you'd have all them out an' you'd be right on the bottom o' the ship, the bottom o' the hold. Then yuh nets would be up above the top after you'd finished haulin'.

'Oh yes, you'd start down the bottom o' the hold. An' another thing you did was have a kid board. When you hauled, that shipped inta the kid between the hold an' the rail. The nets slid down that as you were pullin' 'em in. There'd be about 4 of yuh on deck haulin' in an' pickin' out. You'd throw 'em inta the kid. What you left in, them down below in the hold would be a-pickin' out as the nets come down. You shot the nets just the same as you did for herrin' an' you had a tissot on the warp just the same, but you never had such a thick warp for mackerel nets because they're afloat all the time, which means there isn't so much strain.

'That all depended on the voyage, what sort o' pay-orf you got. Praps one season would be good an' the next bad. That all depended what mackerel there was, because I mean you'd hafta go 100 miles from Land's End sometimes t' git mackerel. Thass a long steam, that is. An' then sometimes you'd git horse mackerel. Scads! Dun't mention them, for goodness sake! You can't shake them out of a net. You can shake an' shake an' shake an' they'll never come out. See, the gills are so hard an' thass the trouble. They give yuh a nasty prick. They're blinkin' terrible things, they are. Mind yuh, you'd more or less know where they were when you were mackerel catchin' by the colour o' the water. You'd just look at the water. That'd be a right whitey colour an' you'd say, "I'll bet there's scads here." An' if you shot in it, nine times out o' 10 you'd git 'em. There isn't much of a sign for mackerel, not out on that deep water. See, you're out on the Atlantic when you're mackerel catchin' an' there isn't much of a sign like there is for herrin'. Chance time you'd praps see a whale, an' thass a good sign, but otherwise there ent much of show. You'd just take yuh chance.

'Like I say, if the water was that lighty colour, that meant scads. So you wouldn't shoot in that if you could help it. O' course, you'd git mackerel in with the scads sometimes. You'd git a mixture. When that happened, you'd pick the mackerel out an' leave the scads in the nets. You'd let 'em die first 'cause they used t' come out better when they were dead. You used t' sell 'em, yuh know. Yeah, you used t' sell 'em when you got inta Newlyn. You used t' sell 'em for bait, for crab pots an' that sort o' thing. You used t' do that with the mackerel sometimes as well — leave 'em in the nets if they didn't come out easy. If they got pulled about too much comin' out, you could sell them as well. You'd git about half-a-crown a cran for 'em. The good ones you sold by the hundred, what they called a long hundred. See, a warp used t' be four, two fish in each hand, an' there were 30 warps t' the long hundred.

'There used t' be several Low'stoft an' Yarmouth boats round at Westward. I spose there'd be about 30 or 40 altogether. The ol' weather could git very bad round there sometimes. I have come inta Mounts Bay (you know, where Newlyn is) an' you couldn't keep the lights alight on board 'cause there was so much wind an' weather. Cor, that used t' be terrible. The water'd be over yuh like a fog. See, that blow right in there — especially when you git a south-west wind. Thass the worst wind comin' in there. That fared better down in the engine room than what that did on deck. You know, you dint git so much o' that rollin' motion. She'd be rollin' a lot, an' you fared better down below than what you did on deck. Mind yuh, there were times when you went t' put a shovelful o' coal on the fire, the boat'd roll, an' you'd finish up acrorss the other side o' the engine room.'

Ernie Armes did not share the East Anglian disdain of the mackerel. He can remember some of the good mackerel landings on the Lowestoft fish market during the 1920s and 30s:

'At one time o' day you used t' git one or two boats go up the Cromer Wold after mackerel. They used t' bring some lovely mackerel hoom from there. Yes, that they did. They used t' git a fair quantity too, only that dint last long. Mackerel catchin' dint last long either spring or autumn; that'd soon finish. Some o' them boats used t' git a last, a last and a half, yuh know. A last was 12,000 fish, 100 long hundreds. See, there was 120 mackerel t' the hundred. That was 30 warps, at 4 fish t' the warp. They moostly went t' Billin'sgate, the mackerel did. They're a nice eatin' fish too. I love a mackerel. Bloomin' lovely! Now a lot o' people wun't touch 'em, but I never minded.

'Durin' the hoom fishin' some o' the boats would praps git some mackerel in among the herrin'. There might be a couple o' baskets. Well, nine times out o' 10 the crew used t' have them for beer money — unless that was a rotten skipper an' he said, "Oh, shove 'em up the sale-ring." They'd hafta t' be sold there then. But moostly the crew used t' have 'em. The skipper would say, "You can have them. Ol' so-and-so will come an' buy 'em orf yuh." That money used t' be shared out amongst the crew and there'd praps be a few pints each for 'em. If they got a lot o' mackerel in with the herrin', they'd hafta go up the sale-ring an' be sold, but not if there was only a few. If they were gittin' mackerel in with the herrin', they'd hafta pick 'em out o' the nets because they couldn't shake 'em out. Sometimes they used t' come ashore in the baskets, in with the herrin'; you know, as the boats were landin'. Praps there'd be a two or three in a basket what the crew hent bothered t' pick out. Well, we'd have them an' that'd be a little beer money for us. You might even git the occasional pilchard too, but not very often. What the poor ol' fishermen didn't like was a net full o' bloomin' scads. They were terrible, they were. There was no shakin' them out. You could eat 'em, but nobody wanted 'em. They were only good t' sell for manure in them days. They kipper 'em now an' they're very nice to eat. O' course, they're nothin' like a bloomin' mackerel; they haven't got no chevrons or nothin' on 'em. They're just got that nasty ol' prickly back. But so have a mackerel come t' that. They can give yuh a jab.'

Scads or no scads, all fishing was hard on the hands. For a last look at the mackerel fishing Ned Mullender casts his mind back more than 50 years and recalls some of the voyages he made.

'When you shot the first net you'd have a white bowl on there on a short strop. Then, when you'd shot about 20, you had what they called the castin' bowl. That was a different colour. Then you'd come t' the quarter bowl, which was a quarter blue an' three-quarters white, or whatever colour the firm painted theirs. (That was t' tell yuh that a quarter o' the fleet was left, when you were haulin'.) Then you'd come t' the half bowl, which was half blue an' half white, see. Then there was the three-quarter bowl an' that'd be three-quarters blue an' only a quarter white. Right at the end you'd have an ordinary bowl agin, praps just a blue one or a black one. The castin' bowl was there t' let you know you were gittin' towards the end o' the fleet when you were haulin', because nine times out o' 10 you hauled round there in the dark. That was a night fishin' moostly, see. You used t' have 250 odd boxes stowed away in yuh fore-hold an' after-well. If you were at sea for two or three days, which you were quite often, you'd put the mackerel in the boxes an' ice 'em. That was very rare you got the full tale (120 fish) in a box round there because they were usually biggish mackerel. You'd git about a half-hundred in; you know, 60 fish. You used t' carry about a ton and a half o' ice an' you'd ice 'em down after you'd picked 'em out o' the nets.

'With the rough nets you shot somewhere about 220, 240, nets. Yeah. You had double the length o' warp that you did for herrin' catchin'. Double the warp. What they did when you went mackerel catchin' was t' take the wings out o' the rope-room so you had that extra space. Then instead o' havin' about a dozen ropes, you'd have 24, praps 25. The rough nets were small; they were only about 10 score deep. An' they were a lot shorter than a herrin' net as well. On each end of 'em there was a big, thick, oblong cork an' that would have yuh boat's name an' number burnt on. Now the rest o' the way along the net-rope the corks were smaller, but still oblong. You used t' have all the hold gutted out t' take the quantity o' nets. I mean, towards the end o' the season, when the mackerel wun't so thick, you sometimes increased yuh fleet t' 14 score o' nets. Thass 280.

'The rough nets were a thicker cotton than the Scotch mackerel ones. They were somethin' like a twine an' they were dressed hard as well. Down on the Westward voyage, because that was deep water, you used t' make what was called a lint foot-rope on the nets. You'd take about eight or 10 mashes up an' beat 'em all the way along. That used t' make 'em hang better in the water. Then you used t' tie 'em together down the sides with little bits o' twine so the fleet hung together. The seizin' came from the net-rope right the way down t' the warp. The Scotch mackerel nets were a lot different; they were much more like a herrin' net, only not such fine cotton an' not so deep. They weren't as good as the rough nets when it came t' pickin' out, though. See, they were finer; an' the finer the nets were, the harder it was t' put the mash over the mackerel's gill without breakin' it. You mustn't have a lot o' broken gills; you couldn't sell 'em like that. Well, you could sell 'em, but you wouldn't git such a good price. When you shot, that was much the same as shootin' herrin' nets, but you dint half notice it when you were gittin' down towards the bottom o' the hold. Cor, they were heavy t' pull up then! Sometimes the corks would heft in the norsels an' that meant that everything would come tight an' pull down the man standin' aft. Yeah, that'd pull him right over. Well, if he had any sense, he'd lay down an' let the net pass over him till the ol' man went full astern.

'One thing about mackerel catchin' is that it's a clean voyage. Herrin' catchin' is a filthy job, with scales an' all sorts o' muck, but mackerel catchin' is a very clean voyage. There'd be about three of yuh on deck pickin' 'em out as you hauled — one on the net-rope an' two on the lint. You used t' pick the mackerel out an' throw 'em inta the kid. In the middle o' the kid you used t' have what they called bank boards. They fitted in at an angle between yuh rail an' yuh coamins, an' thass what yuh nets slid along when you were haulin'. Then, besides you blokes on deck, there'd be two or three more down below stowin' the nets, pickin' out what you dint git an' throwin' 'em inta the fore-room where the boxes were. You dint have no bunker lids orf up on deck; you dint stand in the mackerel like you did with herrin'. You just picked 'em out o' the nets an' threw 'em inta the kid. When the space there was full up, you stowed 'em away under the bank boards. All the nets went down below, unless you were gittin' a real lot. I remember once, we got three lasts, an' I dun't know how many blinkin' scads we dint have with 'em! Well, we stowed a lot o' the nets on deck 'cause we'd filled the hold. All the perk boards were took out an' you started from the bottom. We filled all that space up an' we had the nets up level with the wheelhouse windows time we'd done.

'That was very terrible pickin' 'em out, I can tell yuh! Many o' the chaps got poisoned hands. But then o' course that all depended on how you looked after yuh hands. I mean, a chap what neglected his hands, well, he must expect t' have suffin happen to 'em. I allus used t' carry a bottle o' Izal with me. A bottle o' Izal. Thass a disinfectant. I used t' take a bottle o' that with me, an' when we'd finished haulin' I used t' put a little o' that in some water an' wash my hands in it. If you wun't careful that ol' red bait would soon poison yuh hands. Well, for about a shillin', which was what a bottle o' this stuff corst, you could make sure you were all right. When you got inta Newlyn you used t' take a sample o' the catch up t' the sale-ring. Once you'd sold 'em, you used t' start gittin' 'em out. Then you used t' take 'em up t' the man who'd bought 'em an' shoot 'em out o' yuh boxes or baskets inta his troughs. His blokes used t' wash 'em then, afore they packed 'em away. They used t' ice 'em an' pack 'em away in boxes; then they used t' send 'em orf. Course, where they went to, that was their business, but I spect a lot went t' London. The Newlyn people used t' git a lot o' work by us goin' round there. Oh yis, there wun't any fights or anything like that in my time. The Newlyn Riots were all forgot about.

'That was 1920 when I first went mackerel catchin'. I went 'awseman in the ol' *Silver Herring (LT 1145)*. We went away in March that year, so we were late. That was a late start. There was a lot o' boats ahid o' us what'd got nearly £1000 by the time we got down there. Well, we got down there an' we spent the first day gittin' ready. See, you carried a number o' spare nets an' they'd gotta be put ashore up the net store your firm had hired. All the gear what you dint want, you stored that ashore so you'd got plenty o' room on board. Well, we went out the first night an' I don't think we got more'n about 12 or 14 hundred, so we had another night out an' got about the same agin. Then we come in an' landed. Well, the next night when we went out, there'd bin a boat what'd got a good lot o' mackerel just outside The Wolf. So away we go. We only shot 9 score o' nets 'cause there was a lot o' fish about, an' when I come on watch about 11 o'clock there they were, shootin' around in the water. You could see the fire of 'em; they were right luminous. When we had a look on there was plenty in the nets, so we went t' work. Well, bor, there the boys hung — thick an' heavy. That was slow purchase on that shot too 'cause we had about 3 lasts and a half. When we got in we made £341. I'll allus remember that. Now we were round there at Newlyn from the middle o' March till about the second week in May, when we come hoom. We earnt £2400 for the voyage, which was good. After the expenses had been paid we got our share. I can't remember exactly what it was, but we dint do bad. You only used t' carry a crew o' nine on the mackerel catchin' so that made yuh share better for a start.

'After we'd paid orf we went herrin' catchin' down at Aberdeen. Then we come hoom an' done the hoom fishin'. That took us through till the next year, 1921, when we went orf down t' Newlyn agin. We went away in the February, but it was a bad year. There was a coal strike on for part o' the time an' we lorst a lot o' time through that. O' course, you had a quota o' coal, but that wun't right a lot. By the time we'd done round there we'd only earnt about £1100 or £1200. An' that was afore expenses had come out! Well, we decided t' come hoom an' have a go for mackerel in the North Sea. We went acrorss t' Ostend an' filled the boat right up with coal. We took all the boxes out o' the after-well an' filled that up. We filled up the bunkers an' we took on a deck cargo as well. We had about 30 ton on board, I should think, an' we come out o' Ostend an' shot our nets up in the North Hinder there somewhere. We had a couple o' shots an' we got about 3000 each time. Hoom we come. I forgit what we made, but that wun't a bad price. That was a payin' trip. Out we go agin for another couple or three nights — another 6000 odd. In we come an' got a price. Well, we were doin' all right, gittin' this little fishin' out o' Low'stoft. Out we go agin for another couple o' nights, an' this time we got 7000 or 8000. As we were comin' in that was blowin' a smart ol' breeze, so we put a tow-foresail up. Next thing we knew was the mast had gone over an' pulled the deck up. That'd caught a gust o' wind an' the deck musta bin rotten. The mast dint break; that was the planks in the deck what give. Well, we got the sail orf, stayed the mast t' the rail an' come in. That was the end of our mackerel catchin' because by the time the boat was riddy for sea agin that was time t' go down t' Scotland after herrin'. We went down t' Peterhead an' that was a dud all round. There was no money earnt at all. Chance boat paid its way, but 90% of 'em dint. That was so bad that we runned the voyage inta the hoom fishin' t' try an' git a pay-orf.

'Mind yuh, you got them bad voyages, them bad years. I remember in 1925 I went in the *Welcome Boys (LT 11)* round the Westward, mackerel catchin'. We had the Scotch mackerel nets, the mackerel-herrin'-nets, an' I might say only once did we git any mackerel. We got 12,000 one shot; a last. About 1100 for two shoots was normal that year. We were in debt when we come hoom. The only thing we got was the beatin' money. We used t' mend our own nets an' you'd git what they called beatin' money for that. That wun't very much — about 30 bob t' divide among 5 or 6 of yuh. An' that was for the whole voyage. No, you dint git a lot o' money on mackerel catchin' in my day. Mind yuh, there was one good season directly after the First World War. The first season, in 1919, there was plenty o' boats what got more'n £3000. My brother-in-law, he was in the *Redwald (LT 1171)* an' they got £3300 t' £3400. The skipper's name was Thompson. He was known as 'Golden' Thompson after that voyage.

'The Wolf was a good area for mackerel round at Westward. Then you used t' go down inta the Bristol Channel — not down near Lundy, but orf in the middle. Then another time

COUSIN MACKEREL

you'd be out orf The Bishop. Yeah, about 40 mile south o' The Bishop, out that way. In my opinion, a mackerel is a dirty fish when you git him close in. But out in deep water he's clean; there ent no muck in his gills. He's good eatin' too. You used t' git a gret big thick mackerel orf The Wolf, but down in the Bristol Channel they were thinner. The ones orf The Wolf were full o' oil as well. You know, they were greasy when you cooked 'em. You dint want much fat in the pan when you fried them! You used t' see the oil in the water down there as well. Then there'd be what they called beads. They were like little bubbles on the top o' the water an' they signified mackerel as well as the oil. Orf here, in the North Sea, you used t' git that browny water, the hoss pissy stuff, an' that meant mackerel were about. You got that down at Westward as well, but not nearly so much. Another thing I noticed with mackerel, specially down at Newlyn, was that every now an' agin you'd hear 'em. You know, they'd splash. They'd be swimmin' in a school an' all of a sudden they'd splash an' that'd go right acrorss the school. I don't know whether they done it with their tails, or what they done it with, but that'd run right acrorss the school. That'd let yuh know how many was there, but nine times out o' 10 you dint git none. Not where you could see 'em on the top.

'At one time o' day there used t' be a big spring mackerel voyage out orf here. That was afore my time; I can't even remember it. I can remember the autumn mackerel catchin'. Some o' the sailin' boats went. That was durin' the hoom fishin', October-November time. The ol' sailin' boats used t' go out with the rough nets. That was when I was a young boy. I don't know exactly where they went — out orf here somewhere, I spose. See, I was only a young boy in them days. I'm talkin' about 1907, 1908. I know they used t' git quite a few, though. I can picture 'em now, the ol' sailin' boats. They'd come in an' land at the Waveney Dock, then go an' clean their nets on the concrete. If they'd got scads, they used t' go an' pick 'em out on Sladden's Pier. Oh, they were murder, the scads. When we were round at Newlyn, there allus used t' be several French crabbers in port, an' if we'd got scads we used t' set the Frenchmen on, a-pickin' 'em out. Sometimes, when they'd got what they wanted, they used t' buzz orf. Then you'd git another lot comin' along, wantin' what you'd picked out an' chucked on the deck. "Oh no," we used t' say. "If you want 'em, you come an' pick 'em out o' the nets!"

'I remember one year when I was in the *Hope (LT 1075)* along o' George Rushmore, we got a shot o' mackerel up orf Scarborough. That'd be about the beginnin' o' August an' we were after herrin' really, but we got mackerel. We went inta Scarborough for the weekend an' started sellin' them there on the market. Well, we got rid of about four cran among the buyers, then that dropped away an' we were left with a lot on the deck. The skipper say, "We'll sell 'em t' the visitors, so many on a string." Away we go, six for a shillin', an' we did all right. This was 1919, the first year after the war. In the end we were lettin' 'em go for a penny each, just t' git rid of 'em. There was people comin' an' goin', an' chance we'd have a hawker come an' buy a basket, then buzz orf somewhere. I can remember one woman comin' up to me. She say, "What are they?" I say, "Mackerel, ol' dear." She say, "I know that, yuh fule. I mean, how much are they?" I say, "Well, why dint you say so? They're 12 for a shillin'. Penny each." So she had a shillin's worth an' orf she went. We sold nearly all of 'em that day, an' when we doled up we had somewhere round about 35 bob each. There was 10 in the crew, so that wun't a bad little trip.

'When you first went round t' Newlyn the March mackerel were a big, thick mackerel. They had a lovely roe in 'em an' they were thick, round an' stubby as regards the build. The mackerel you used t' git orf here about the same time were a long, thinnish mackerel, in my experience. I dun't know whether they had a roe in 'em or not 'cause I never did eat 'em. The autumn ones were lovely fish, though. Yes, they were a good class o' mackerel, they were. I never ate a lot o' mackerel round here. I used t' like the Westward ones. I mean, I've even eaten mackerel's liver round there. You used t' have that raw, with pepper, salt an' vinegar. You used t' do 'em like you did a whilk or a cockle an' have an enamel mug full. Cor, they used t' be lovely. That was a craze o' the crew t' have them like that. You could fry 'em as well, the livers, an' they were lovely like that too. I don't think you can beat a nice mackerel from the west coast, round about March time. Well, when you come t' weigh it up, there aren't many fish in the sea what aren't worth eating.'

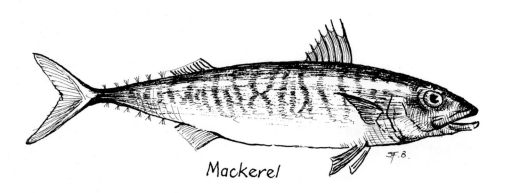

Mackerel

Typical Lowestoft fish merchants premises and yard. This belonged to Tom Brown & Son Ltd and was at the bottom of Maltsters Score. Below, a good shot of herring being discharged. Unfortunately big catches did not always mean high prices and many hours of hauling nets often resulted in a poor return.

The End of The Voyage

'The farmer has his rent to pay,
Haul, you joskins, haul,
And seed to buy, I've heard him say,
Haul, you joskins, haul,
But we who plough the North Sea deep,
Though never sowing, always reap
The harvest that to all is free,
And Gorleston Light is home for me,
Haul, you joskins, haul.
           (Traditional — A Yarmouth Shanty)

The lines above fittingly conclude this history of the East Coast herring industry and admirably sum it all up. First of all, there are the joskins, the country lads who left their farms after the harvest was over and came down to Yarmouth and Lowestoft for the autumn fishing. Things were slack then on the land, and as the herring boats got bigger round about 1810 or so their muscle was needed to power the hand-capstans when hauling the nets. The ever-increasing expansion of the industry throughout the 19th century drew more and more of them down to the coast, until by the time that the steam drifter came along country fishermen were a well-established fact — and not just on a seasonal basis. Many of them now followed the herring the whole year round, walking and biking down to Yarmouth and Lowestoft at first, and using buses, trains and taxis in later years. Jumbo Fiske, Jack Sturman, Horace Thrower and Frank Fisk were all originally from out of town.

Then there is the emphasis in the refrain on the sheer physical effort involved in fishing. This continued long after hand-capstans had disappeared and all crew members on board a drifter knew what it was to work. The refrain echoes through these pages too. Memories of a life of unremitting toil for little or no reward. The 1920s and 30s were every bit as hard for the fisherman as they were for the industrial worker, maybe even harder, and any fortunes made during that time didn't go to those who steamed out in all weathers in pursuit of a precarious livelihood. The few men who prospered usually did so by capitalising on the misfortune of others, buying up boats and bankrupt businesses at rock bottom prices and then hanging on till better times.

Yet no matter what the economic state of the industry, one belief remained constant among all the people engaged in herring fishing: the conviction that the shoals would always be there for the catching. "The harvest that to all is free" — that line from the shanty says so much about attitudes to the herring. There for the taking; there for ever; there to be dumped even, if the price dropped too low, because there were plenty more in the sea. And all this was true within the context of the time because there was no reason to think otherwise. Even so, it was a free harvest that was bought dear more often than not. There were the long voyages which saw no profit at the end; there were the blistered hands and saltwater boils as everyday discomforts; there were accidents on board that resulted in temporary or permanent disability; and for many fishermen there was, ultimately, a final resting place beneath the waves.

So our witnesses in this oral history of the 20th century East Anglian herring fishing have reached the end of their evidence. It has not been a complete or definitive statement, and there was no intention that it should be so. The whole purpose was to set down the

THE END OF THE VOYAGE

experiences of people who were vitally and directly concerned with the fishing, and who saw both the end of the herring and of the way of life associated with it. Well perhaps not the whole purpose. The sheer enjoyment of their yarns was pretty important too.

It must be doubtful whether a large-scale commercial fishery for the "silver darlings" will ever come back to either Yarmouth or Lowestoft, and even if it does then it will probably not be based on drift netting. At the time of writing there is a total ban on fishing for herring in the North Sea, a restriction that is regarded everywhere locally with a mixture of frustration and disbelief. After all, it isn't the East Anglian longshoremen who have caused the depletion of stocks, so why should they be deprived of a traditional and seasonal income? Perhaps they are not being entirely cut off from the autumn shoals. The fisherman is a hunter and the hunting instinct is strong. For centuries he has been adept at snatching a living from the sea against every adversity. If an occasional quarter cran is still being landed around the time of the harvest moon and the "hoom fishing", I should not be surprised.

The "Home Fishing" - Lowestoft drifters off to sea again in the 1930's.

Glossary

After-well	fish-hold between the main hold and the wheelhouse
Agin the law	term used of the wind getting round and blowing from the opposite direction.
Allotment	weekly payment made to drifter crews (deductable from their pay-off).
Back or back-rope	the double rope at the bottom of a herring net
Bale	to pull up a herring net into a bag, when hauling, so as to stop the fish from falling out.
Bank board	a board that fitted in between the boat's rail and the hatch coamings, down which mackerel nets were hauled in.
Bark	to tan drift nets
Barmskin	oilskin
Bass	a very heavy rope
Beat	to mend drift nets
Beatster	a woman that did this
Beckett	a loop in the end of a rope
Beef-kettle	a large cast-iron cooking pot
Bend	a way of joining 2 ropes without splicing them
Berth	a bunk, or a boat's mooring space, or a job on board ship
Bethel	a nonconformist chapel of no set denomination
Bloater	a herring lightly salted and smoked with the gut in
Blow down	drop steam pressure and extinguish the boiler fires
Blow off	build up steam pressure until the safety valve released it
Blower	a whale or a porpoise
Boom	the spar on the foot of a sail
Bowl	a wooden cask that acted as a float on herring or mackerel nets
Braid	to make trawl nets
Braider	someone who did this
Buff	large canvas or plastic float on herring nets
Bunker	below-deck coal storage space
Bunker-lid	manhole cover in the deck of a fishing boat
Buss	a large 16th/17th century herring boat
Capstan	upright cylindrical winch for winding in ropes
Carvel-built	boat hulls constructed with the planks flush
Cast-off	crew member who untied the seizings from the warp and regulated the speed of the capstan when hauling
Chittled	term used for when drift nets rolled up on themselves
Chuck up	term used of a buyer refusing to accept a catch of herring
Clinkerbuilt	boat hulls constructed with the planks overlapping
Close	dusk
Coalies	coalfish, or saithe
Coamings	timbers round the mouth of the hold on to which the hatches batten

GLOSSARY

Coble	the Yorkshire longshore boat
Cod end	the bag at the end of a trawl net, which holds the fish
Converter smacks	sailing boats that went both drifting and trawling.
Cran	a measure of 37½ Imperial gallons; 28 stones of herring by weight
Cran out	unload herring from a drifter
Crotch boots	thigh boots
Crow's foot	two strands of a mesh broken
Cutch	substance used to preserve drift nets
Dan	a kind of marker buoy
Dandy	a ketch-rigged sailing drifter
Denes	the grass or heath area between the cliffs and the beach
Didall	a round net on a long pole, rather like a giant butterfly net
Dodge	keep a boat head to wind
Dogs	dogfish
Dog-eaten	term used of drift nets damaged by dogfish
Dons	top skippers
Double-swim	herring enmeshed on either side of the net
Draw-bucket	pail used to get water from over the side of the boat
Drift nets	nets that hang in the water and catch fish by their gills
Drifter-trawlers	steam boats that did both kinds of fishing, according to the season
Driver	chief engineer on a drifter
Drive down (and up)	term used of a boat drifting on the tide behind its nets
Drop on	to come across herring
Duff-chokers	nickname for Yarmouth fishermen
Dump heads	ends of the piers at the dock entrance
Eye	loop in the end of a rope
Fang	propeller blade
Fireman	Yarmouth term for stoker
Fish room	main hold for the catch
Flakes	the serpentine coils of a rope
Fleet	a set of nets
Flews	herring nets, a general term in the mediaeval period
Foc'sle	living quarters in the bow section of a boat
Foc'sle funnel	flue pipe from the foc'sle stove
Foot-rope	the back-rope
Forelock	loop in the end of a drift net's heading cords
Forestay	the wire support of the foremast, leading down to the stem
Gaff	the spar on the head of a sail
Garden lint	herring nets so badly damaged as to be no use for anything else
Gill	the side of a herring net down the heading cords
Gilldeds	herring still enmeshed after scudding

GLOSSARY

Gland	metal sleeve that kept the packing round a piston tight and in place
Going off	tending a drifter's engine on the first part of an outward journey
Golden herring	medium salted and smoked fish
Gorger	a large enamel jug
Gross tonnage	The term has nothing to do with weight. One gross ton simply equals 100 cubic feet of enclosed space below deck. Net tonnage is the gross figure less the boat's non-earning spaces, such as the crew's living quarters, galley, etc. In the steamboats the engine-room was an important non-earning space.
Hawseman	the man below mate on a drifter
Headings	the cords down the side of a drift net
Heft	catch, come fast
Home fishing	the East Anglian autumn herring season
Hoodway	the entrance to the cabin on a boat
Improver	a beatster in her second year
Inside	close to the shore
Jellies	jellyfish
Joskins	countrymen who went fishing
Jumper	a calico or cotton slop
Kid	the space on a drifter's deck between the hold and the gunwale
Kid board	see bank board
King Herring	the shad
Kipper	a herring that is split, salted and smoked
Klondyke herring	fish exported to Germany in salt and ice
Klondyke boxes	large wooden cases for shipping the above, holding 12¼ stones
Last	a measure of herring or mackerel (100 long hundreds q.v.)
Light duff	dumplings
Lining	long-lining
Lint	the meshes of a net
Little boat	a drifter's dinghy
Long hundred	120 mackerel, 132 herring
Long-jawed	term used of a rope coiled the wrong way, which pulled the strands out of line
Longshore boat	one that fishes close in (literally 'along the shore')
Look on	haul in a net or two to see if there were herring or mackerel in
Lugger	a boat with lugsail rig
Lugsails	large, square sails set on a yard. A transitional design between square and fore and aft rig.
Maise	herring milt or roe. Maisy herring were those full of milt or roe, about to spawn.
Make up	to clear up the boat and tidy the gear at the end of a voyage

GLOSSARY

Mand	1,000 fish originally. Later, a large wicker basket for handling sprats.
Manfare	a mediaeval herring net capable of holding a mand of fish (1,000 in number)
Marlin spike	pointed instrument used in splicing ropes
Mizzen	the mast and/or sail aft
Mole-jenny	a portable fair-lead through which the warp ran
Molgogger	see above
Monkey	a chequered buff near the end of a fleet of herring nets
Needle	instrument used for mending nets
Net-rope	the double rope on the top of a drift net to which the corks were attached
Night-man	some-one who stepped in at short notice to take the place of a sick crew member
Norsels	lengths of twine that joined the lint to the net-rope and retained the corks in position
Oddie	the guarding meshes at the top and bottom of a drift net
Oily frock	oilskin smock
Ossels	norsels
Overdays	second day herring
Over-overs	third day herring
Part	break (used of warps and nets)
Pay-off	the settling at the end of a voyage
Pea bellies	nickname for Lowestoft fishermen
Perks	boards halfway down the hold on which the nets were stacked
Pick	boathook
Pole-end	the first net cast over
Pounds	compartments for the storage of fish
Pound boards	planks that divided the fish-hold up into separate compartments
Prime fish	sole, turbot, brill, etc
Qualmy	queasy
Rail	top part of the gunwale
Ransacker	man who checked over the nets
Red bait	substance in a mackerel's gut (the copepods on which they feed)
Red herring	heavily salted and long smoked fish
Rednecks	nickname for Yarmouth fishermen
Rigger	man who set up the nets and got them ready for use
Roads	the sea approaches to Lowestoft and Yarmouth
Roaring shovel	wooden shovel used for moving herring
Rope-room	the space below decks where the warps lay coiled
Rough net	a drift net made of hemp twine or heavy cotton

GLOSSARY

Running a trial	taking a new boat out on her trial trip before the owners took delivery from the yard
Run down	to clean drift nets and stack them neatly
Runtie	a Scottish sailing drifter
Saturday and Sunday night boat	a vessel that laid in port and didn't fish on Sundays.
Scad	the horse mackerel
Scandalise	slacken the mizzen sail by dropping the peak of the gaff
Scotch cure	herrings pickled in barrels of brine
Scran	inferior herring, undersized and damaged
Scranners	small buyers of odds and ends
Scud	shake herring out of the nets
Scummers	herring that had fallen from the nets and been caught in a didall
Scutcher	a metal or wooden scoop for loading herring into baskets (from the Latin 'scuta', meaning a shield)
Seizing	the rope that joined the bottom of a herring net to the warp.
Set in	to rig drift nets so that the meshes were diamond shaped
Sheer water	clear water
Shimmer	a quantity of herring in one haul, probably derived from the silvery sheen of the herring which is due to a substance called guanine
Ship's husband	the man in charge of crewing arrangements and the boat's well-being
Shoot	to cast nets
Shot	the quantity of herring caught at any one time
Sidecords	the twine along the top and bottom of a herring net
Slice	long metal rod for pushing coal down into bunkers
Smack	a ketch-rigged sailing trawler
Speets	wooden rods on which fish are hung to cure
Spell buffs	floats that marked the quarter stages of a fleet of nets
Spell round	change over jobs when hauling nets
Spents	herring that had shot their roe
Splice	join 2 ropes together by unravelling and replaiting the strands
Split-arses	dumplings that had burst open (also vulgarly applied to nuns)
Spoilt	a damaged drift net
Spronk	one strand of a mesh broken
Stanchions	the vertical supports in the framework of a boat's hull
Stem	the bow post of a boat
Stem end	the net nearest to the drifter
Stern post	the upright timber at the rear end of a boat
Stocker-bait	money made by the fishing crews over and above their wages (done by selling surplus fish, net cleanings, etc)
Stockie	as above

GLOSSARY

Strop	the rope which joined the top of a drift net to the buff
Swale round	term used of nets sweeping round, back in towards the boat
Swim up	term used of herring rising
Swing	the part of the warp between the boat and the nets, usually made of bass
Tan	to preserve drift nets by boiling in cutch
Tanning copper	large brick tank in which nets were boiled
Tanning tank	metal tank in which nets were preserved on board ship
Thief net	a poke net that hung under the drift nets and caught any herring that fell out
Third hand	man below mate on a trawler
Thrush	coupling that connected the engine to the crankshaft
Tie	a cord put round a drift net which had been folded up and put aside for repair
Tissot	a strong rope that took the strain of a fleet of drift nets
Tonnage	(see Gross Tonnage)
Tow-foresail/tow-dinger	a large triangular sail
Trimmer	fishermen on board a trawler or drifter-trawler who prepared coal for the furnace and tended the boat's lights
Trick out	to spread a drift net out on the ground
Turn	the twist of a rope's strands
Vatted herring	fish steeped in brine in under-floor tanks
Warp	(1) master-rope that holds and supports fishing nets
	(2) 4 herring or mackerel
Whaleman	one of the three-quarter and half-quarter sharemen on a drifter
Wherry	a flat-bottomed barge
White elephants	boats that didn't pay their way
Wings	spaces below deck between the hold and the sides of the boat
Woodbine funnel	long, narrow funnel above Elliott pot engine
Wrapper	silk neckerchief
Yarco	a Yarmouth man
Younker	Yarmouth term for the cast-off.

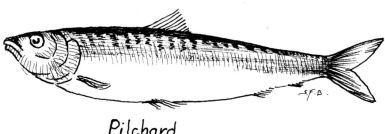

Pilchard

Select Bibliography

An Etymological and Comparative Glossary of the Dialect and Provincialisms of East Anglia by J.G. Nall (Longmans, Green, Reader & Dyer, 1866).

Great Yarmouth and Lowestoft by J.G. Nall (Longmans, Green, Reader & Dyer, 1866)
The East Anglian, volumes 3 & 4 (Samuel Tymms, 1869)
The Herring and the Herring Fishery by J.W. de Caux (Hamilton Adams & Co., 1881)
The Victoria County History of Suffolk (Archibald Constable & Co. Ltd., 1907)
The Herring by A.M. Samuel (John Murray, 1918)
Herring and the Herring Fisheries by J.T. Jenkins (P.S. King & Son Ltd., 1927)
The Fish Gate by Michael Graham (Faber & Faber, 1943)
Sailing Drifters by Edgar J. March (Percival Marshall & Co. Ltd., 1952 reprinted by David & Charles 1969)

The Herring and Its Fishery by W.C. Hodgson (Routledge & Kegan Paul, 1957)
80 Years of Shipbuilding by L.E. Richards (Richards Ironworks Ltd., 1956)
The Roaring Boys of Suffolk by P. Cherry & T. Westgate (Brett Valley Publications, 1970).

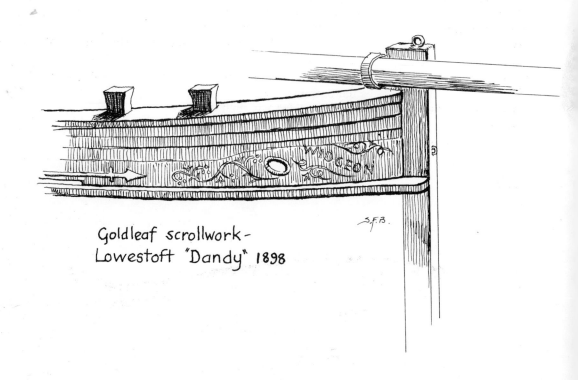

Goldleaf scrollwork – Lowestoft "Dandy" 1898